Social Network Analysis

Scrivener Publishing
100 Cummings Center, Suite 541J
Beverly, MA 01915-6106

Publishers at Scrivener
Martin Scrivener (martin@scrivenerpublishing.com)
Phillip Carmical (pcarmical@scrivenerpublishing.com)

Social Network Analysis

Theory and Applications

Edited by
Mohammad Gouse Galety
Chiai Al Atroshi
Bunil Kumar Balabantaray
and
Sachi Nandan Mohanty

Scrivener Publishing

WILEY

This edition first published 2022 by John Wiley & Sons, Inc., 111 River Street, Hoboken, NJ 07030, USA and Scrivener Publishing LLC, 100 Cummings Center, Suite 541J, Beverly, MA 01915, USA

For more information about Scrivener publications please visit www.scrivenerpublishing.com.

Wiley Global Headquarters
111 River Street, Hoboken, NJ 07030, USA

For details of our global editorial offices, customer services, and more information about Wiley products visit us at www.wiley.com.

Library of Congress Cataloging-in-Publication Data

ISBN 978-1-119-83623-0

Cover image: Pixabay.Com
Cover design by Russell Richardson

Set in size of 11pt and Minion Pro by Manila Typesetting Company, Makati, Philippines

Printed in the USA

10 9 8 7 6 5 4 3 2 1

Contents

Preface

By helping students envision the future, a teacher can help them prepare for it. On this transcendent note, we deigned this book to encourage students to take advantage of the possibilities and opportunities presented in the field of social networking. Several books have been written on the inexhaustible theme of Social Network Analysis over the last few decades. However, this book is a cumulative review of the new trends and applications manifested in areas of social networking.

Our intention was to present an agglomeration of diverse themes of social networking analysis such as an introduction to Python for social networks analysis; handling real-world network datasets; the cascading behavioral pattern of social network users; social network structure and data analysis in healthcare; and a pragmatic analysis of the social web. Also presented are components of Semantic Web mining; classification of normal and anomalous activities in a network by cascading C4.5 decision tree and K-means clustering algorithms; a machine learning approach to forecast words in social media; a sentiment analysis-based extraction of real-time social media information from Twitter using natural language processing; and using cascading behavior in concepts and models to explore and analyze real-world social networking datasets.

We were delighted to see that many authors traversing many realms chose to contribute to this book. The topics covered are categorized according to themes. Chapter 1 discusses the hypothesis of social network analysis (SNA), with a short prologue to graph hypothesis and data spread. It projects the role of Python in SNA, followed up by building and suggesting informal communities from genuine pandas and text-based datasets. Chapter 2 accords with graph representation, Network-X, scope of Python in SNA, and the installation and working environment of Python. Chapter 3 presents the basic principles of scale-free network and its primary scenarios for modeling and analyzing the performance of the network to provide an approximate data from a massive network such as social media. Chapter 4 deliberates the cascading behavioral pattern of social network users with the user-generated content consisting of images, text and videos. Machine learning algorithms and natural language processing help to understand the text content of data and the user behavioral pattern in social media. Chapter 5 develops a deep insight into SNA and its applications in the healthcare system.

Continuing on, Chapter 6 proposes an integrated model approach with social semantic ontology under a specific (agricultural) domain which is composed of domain ontology and social ontology. This integrated approach is used for establishing social semantic ontology. Chapter 7 elaborates the method of identification of anomalies with "K-means + C4.5," the method of cascading K-means clustering and the C4.5 decision-tree methods for classifying anomalous and typical computer network operations. Chapter 8 establishes forecasting as one of the machine learning and supervised learning algorithms. It builds models that capture or explain the data to figure out the reason for the fundamental causes of a time series through a term frequency and inverse document frequency algorithm. Chapter 9 presents a machine learning algorithm using Naïve Bayes method that analyzes polarity in twitter streams. Sentiment analysis is effective in mining sentences taken from Twitter. Chapter 10 deciphers cascading behavior, and discusses its purpose and significance with special focus on decision-based, probabilistic, independent cascade, linear threshold and SIR models. The concept of centrality, cascading failure and cascading capacity are also elucidated. Chapter 11 devises a Python framework for analyzing the structural dynamics and functions of complex networks.

We sincerely believe that this book will prove to be a useful augmentation to Social Network Analysis. We would like to express our appreciation to the authors, publisher and the team members for their strenuous efforts. Lastly, we thank our family members for their support, encouragement and patience during the entire period of this work.

Dr. Mohammad Gouse Galety
Mr. Chiai Al-Atroshi
Dr. Bunil Kumar Balabantaray
Dr. Sachi Nandan Mohanty
March 2022

1

Overview of Social Network Analysis and Different Graph File Formats

Abhishek B.[1]* **and Sumit Hirve**[2]

[1]*Department of Mechanical Engineering, University of Applied Sciences, Emden Leer, Germany*
[2]*Department of Computer Engineering, College of Engineering Pune, Pune, India*

Abstract
Evaluating the public data from person-to-person communication destinations through the social network could create invigorating outcomes and bits of knowledge on the general assessment of practically any product, administration, or conduct. One of the best and precise public notion pointers is through information mining from social networks, as numerous clients seem to state their viewpoints on the social networks. The innovation in the Internet technologies figured out how to expand action in contributing to a blog, labeling, posting, and online informal communication. Therefore, individuals are beginning to develop keen on mining these immense information assets to evaluate the viewpoints. The Social Network Analysis (SNA) is the way toward researching social designs using graph hypothesis and networks. It integrates an assortment of procedures for analyzing the design of informal organizations, in addition with the hypotheses that target describing the hidden elements and the patterns in this framework. It is an intrinsically integrative field, which initially emerged from the sectors of graph hypothesis, statistics, and sociopsychology. This chapter will cover the hypothesis of SNA, with a short prologue to graph hypothesis and data spread. Then discuss the role of Python in SNA, followed up by building and suggesting informal communities from genuine Pandas and text-based data sets.

Keywords: Data mining, SNA, viewpoint dynamics, graph hypothesis, Python

**Corresponding author*: abhishek.dilip.bhambere@gmail.com

Mohammad Gouse Galety, Chiai Al Atroshi, Bunil Kumar Balabantaray and Sachi Nandan Mohanty (eds.) Social Network Analysis: Theory and Applications, (1–18) © 2022 Scrivener Publishing LLC

1.1 Introduction—Social Network Analysis

A network of interactions, where the nodes comprise of number of people, and the edges comprise of interaction among the people are termed as social network [1]. The numbers of social networks and the strategies to analyze them are available since the past decades [2]. Statistics, graph theory, and sociology are the basics for the development of the area of social networks and are used in number of fields, such as business, economy, and information science [3, 4]. The analysis of a social network is analogous to the analysis of a graph because of the presence of graph, like topology of the social network. Graph analysis consists of a number of strategies but is not suitable to analyze the social networks [5–7] because of its complex characteristics. A very large-sized social network comprises of millions of edges and nodes, where the node generally possess number of attributes. The complex and large graph of social network cannot be managed using the old graph analysis strategies [8].

Email network, collaboration network, and telephone network are the various types of social networks. However, recent online social networks, like Twitter, Facebook, and LinkedIn, have gained increased popularity within a short period with a greater number of users. It was found with a survey that Facebook has crossed more than 500 million users in the year 2010 [8]. Social media acts as a highly recognized platform with rich source of data assisting well in the field of marketing of various brands, responding to changes in marketing, enhancing the brands through promotion, and eventually attaining a large number of customers [9–11]. In particular, the role of social network is very important in the area of healthcare applications. As such, the healthcare sector requires discovering new traditions to control the provider practice and measure the best practices to satisfy and improve the health outcomes. Social network analysis (SNA) concentrates on evaluating the relation among individuals, who are attached by one or more knot of interdependency, like friendship, love, trust, cooperation, or communication. Social network analysis can provide imminent into evaluating and understanding the specialized networks of communication and, hence, developing effective interventions in the network to enhance the performance of the provider and eventually, the outcomes related to health [12]. The diagrammatic representation of SNA is shown in Figure 1.1.

For illustration, let us consider that the application of online social network in analyzing the contagious diseases originated with the biological pathogens, such as influenza, chickenpox, measles, and the sexually spread viruses that transfer from one person to another [13–15].

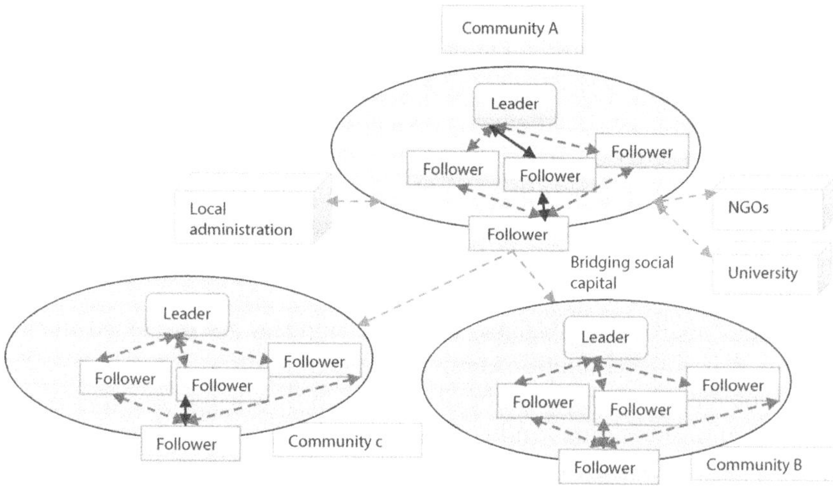

Figure 1.1 Social network analysis.

Recent studies have observed the prologue of a number of SNA models that try to clarify how opinions develop in a population [16], with the consideration of a number of social theories. These models possess a number of common characteristics with that of the spreading and epidemics. Generally, people are considered as agents with a certain state and attached by a social network. The social links is indicated using a complete graph or with more sensible complex networks. The state of the node is typically identified using the variables, which can either be discrete or continuous, with the probability to select either one or another option [17]. The nature of individuals varies with respect to time, depending on a number of update rules, mainly with the interaction of neighbors.

1.2 Important Tools for the Collection and Analysis of Online Network Data

In the recent years, the SNA has attained more concentration in various fields of research, which is because of the flexibility in operation provided by the graph theory that is involved in reducing the countless phenomena to a basic analytical form in terms of bricks and nodes. Certainly, the social relations, transportation, trading, communication strategies, and even the brain can be framed as a network and can be analyzed. This assists in the visibility of the studies related to network analysis, leading to be advantageous in education centers, academies, and universities particularly, healthcare. A number of tools were developed to make it available to a large amount of people. The SNA library and the graphical tools are made available to physicists, mathematicians, computer scientists, and so on. The SNA, being an active area of research, can also be used for unfolding human interactions and opinion diffusions. More number of dedicated tools and libraries are available even for certain peculiar applications. However, it is a time-consuming process to select the appropriate tool for a particular task, making it inconvenient for the users.

Some of the openly available tools and libraries are discussed in this section. A multilevel solution aiming on epidemic spreading simulation is represented as Network diffusion library (NDlib), which possesses a number of significant features and is available highly to the SNA practitioners as compared with other tools. Unlike other tools, the NDlib tool is accessible to technicians, like researchers, programmers, and to non-technicians, like students and analysts. NDlib helps in rectifying the drawbacks associated with the existing libraries with reduced complexity in usage. The three elements of the generic diffusion process are the graph topology, the diffusion model, and the configuration of the model.

The configuration of the model is devised in such a way to provide the final user with negligible and logical interface to choose the diffusion processes. The simulation configuration interface finally permits the user to completely indicate the three different groups of data, such as the model-specific parameters, the attributes of nodes and edges, and finally, the preliminary condition of the epidemic process. The configuration model has an important role in library logic in such a way that it concentrates on the description of the experiment, thus leading the definition of the simulation logical over all the models [18]. The next significant software package is the NetworKit [19], which generally provide the graph algorithms, and is efficient in analyzing the capabilities of the network. It involves balancing certain combination of strength with its two-layer hybrid feature aware code [12]. Figure 1.2 illustrates the SNA using Python.

Social Network Importer: The SN organization is a module for NodeXL6, which is the unrestrained Excel 2010/2007 format for dissecting organization in the well-known Excel application software circumstance. The Bernie Hogan of Oxford Internet Institute delineates the NameGen7, which is considered as the antecedent of SN organization [20].

Social Network Organization Importer: SN organization makes inquiries to Facebook Administration Programming Interface (API) and permits the extortion of inner self-organization information for

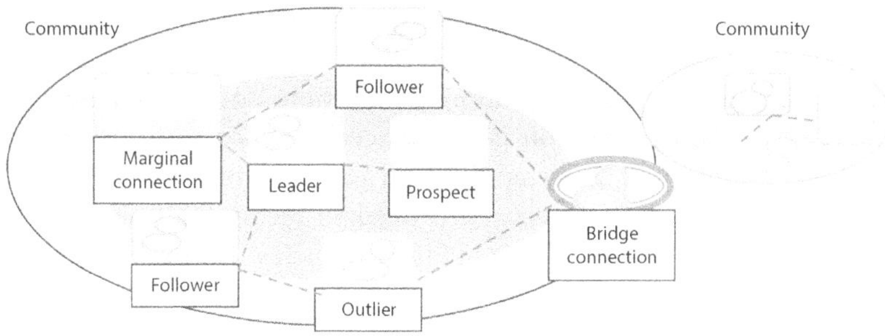

Figure 1.2 Social network analysis using Python.

a provided Facebook client. Contingent upon account protection settings for conscience and revamp, the apparatus will likewise gather Facebook portrait information and restore the 1.5 degree sense of self-organization. As per the Facebook API protocols and regulation, the information must be gathered for a conscience who has given their Facebook username and secret word, and henceforth Social Network Importer is as of now basically valuable for analysts who need to gather their own inner self-organization information or that of few members who might have to utilize NodeXL on a machine that influence scientific approaches. In contradiction, NameGen is accessible as an application of Facebook, and it has permitted the designers of NameGen to gather a sense of self-organization information for individuals who assented to take part in the evaluation, where the assent was conceded by means of the establishment and utilization of the NameGen Facebook implementation. Although the SN Importer effectively conceals the interaction between the researcher and the Facebook API, the Tweepy Python library established for Twitter API is significantly more truncated level in that its utilization requires the specialist to have the option to program in Python [21]. Common utilization of Tweepy may include the specialist questioning the Twitter Search API to track down all new tweets that consist of a specific hashtag.

The API of the twitter clients is then utilized to accumulate the administered assistant network of the writer of the tweets. The Communal Online SM observatory Observant (COSMOS) organization that contributes a consolidated set of devices for gathering, documenting, exploring, and envisage the data streams in the social network, along with the ability to connect with the variant types of data, such as the data from UK ONS (organization of national statistics) through the extended APIs [22, 23].

The COSMOS holds a scope of demographic devises which comprised of gender recognition, stress, topic realization, language identification, location identification, and emotion recognition. The initial description of the COSMOS organization is being accessible for transfer from the late 2014. The Python Flickr application gadgets are delineated for the Python software programmers, who need to technologically interconnect with photo distribution sites of Flickr websites. The experimentation make utilization of the Python Flickr API, which might involve acquiring Meta information, such as descriptive tags on the flicker images transferred through the specific Flicker participants then, at that point, repeating over directory of description data and establishing a semantics network at where the suspended and biased tie between labels, determines the measure of times that were conjointly utilized to portray a particular photograph. At long last, the VOSON apparatus for interface network grouping and evaluation is accessible as each web application and module to NodeXL.9 users will enter a posting of seed URLs (regularly, passage pages to net sites), and furthermore, the web crawler would then be able to creep through each site and gather active text content and hyperlinks. Alternatively, the crawler comes showing up hyperlinks to one and every page inside the site (this is as of now accomplished by means of the VOSON code getting to the Blekko net PC program API10). VOSON grants the client to develop organizations of web substance or sites, and these are frequently imagined inside the net application and its capability to move networks for investigation in elective organization examination instruments.

- **NodeXL** (http://nodexl.codeplex.com) is characterized above with regard to information assortment. However, it additionally gives a menu-driven circumstance to organize perception and examination.
- **Pajek** (http://pajek.imfm.si/doku.php) is a Windows-dependent catalog-operated collection of data, recognized for its capacity to deal with enormous organizations. Pajek is the broadly utilized system Software for designing the organizations, Pajek has insightful capacities, and can be utilized to process most centrality measures, recognize primary openings, block model, and so on. IGraph is a free programming package for making and controlling charts. It incorporates executions for exemplary diagram hypothesis issues like least crossing trees and organization stream and, furthermore, carries out calculations like local area

structure search. The effective execution of IGraph permits it to deal with diagrams with an enormous number of edges and nodes.

- **Statnet** (http://statnet.csde.washington.edu) is a subset of R, which is an extended source factual programming library for organization administration and examination, incorporated with ERGM.
- **NetworkX** (http://networkx.github.io) is one of the Python language programming packages utilized for the network evaluation. x'x'. Networkx is the Python language programming packages for the formation, exploitation, and evaluation of construction and elements of the unpredictable organizations. With the support of this apparatus, the user can deliver and reserve the networks in the recognized information designs, can create numerous kinds of arbitrary and exemplary organizations, dissect network structure, construct network models, draw organizations, and so on. Networkx has numerous highlights like multIGraphs, language information structures for diagrams, and dIGraphs [24]. Hubs can detain "anything," such as pictures and text, Edges can detain discretionary information, such as loads, time-arrangement, Standard diagram calculations, Network construction, evolutionary measures, and so forth.
- **Gephi** is an intelligent representation and observation stage for a wide range of organizations, dynamic, and various leveled charts. Linux, operates on Mac OS X, and Windows. Gephi are the device for individuals that need to investigate and observe diagrams. Similar to Photoshop, yet for information, the client interfaces with the characterization and control the designs, shapes, and shadings to uncover the concealed properties.
- **IGraph** (http://igraph.org) can be established as the libraries for R, C, Ruby, and Python [4]. More than four instruments are analyzed on the accompanying six measure stage, such as algorithm time intricacy, types of graphs, chart design, diagram input folder design, diagram features, and database for the SNA apparatuses examinations: Slashdot data set is widely accepted data set. It consists of 982787 edges (administered) and 77317 nodes. Slashdot is an innovation related news site that highlights client

submitted and assessed reports about science and innovation related themes. IGraph is a library for network examination that runs in both Python and R.

- **Gephi** (https://gephi.org) executes on Mac OS, Linux, and Windows and is a catalog-operated organization representation apparatus.
- **PNet** (http://sna.unimelb.edu.au) is a catalog-operated Windows collection for ERGM.
- **UCInet** (https://sites.google.com/site/ucinetsoftware/home) is a catalog-operated Windows collection for the SNA [25].

A. Correlation Based on Platform Social organization: The evaluation devices, such as Pajek and Gephi, remains as the solitary programming, which consists of IGraph and Networkx as the libraries. Pajek and Gephi execute on Windows stages where Networkx makes use of Python library, and IGraph makes use of python/c/r library for interpersonal organization evaluation. IGraph, Pajek, or Networkx can deal above 1,000,000 hubs, and Gephi can deal with 150,000 hubs.

Evaluation Based on Network Category: In the SN analysis, there are four kinds of organization graph [26]. In a one-mode organization, every vertex can be identified with another vertex. In a one-mode network, the clients have just one group of nodes, and the restrictions are associated with these hubs. In a two-mode organization, vertices are partitioned into two sets and vertices must be identified with vertices in the other set. Two-mode network Graph are a specific sort of organizations with two arrangements of nodes, and the ties are just settled between the nodes having a place with various sets. Methods for dissecting one-mode networks cannot generally be applied to two-mode networks without alteration or change of significance. Extraordinary methods for two-mode networks are extremely confounded. We can make two 1-mode networks from a two-mode network. In a multisocial organization, there will be different sorts of relations between hubs. Hubs might be intently connected in one social organization, yet far off in another. In worldly organizations (dynamic diagrams), organizations can change after some time. The lines and vertices in a worldly organization ought to fulfill the consistency condition: in the event that a line is dynamic in time t, additionally, its end-vertices are dynamic in time t. For one-mode or two-mode network investigation, we can utilize any of programming apparatuses; however, for multisocial organization chart, we have just Pajek programming instruments; for brief network diagram, we have Networkx and Pajek devices.

1.3 More on the Python Libraries and Associated Packages

The aforementioned libraries are not the main library intended to show, recreate, and study diffusive elements on complex organizations. To all the more likely edge our library inside the arrangement of existing scientific devices, we recognized the following accompanying contenders:

- **Epigrass:** Epigrass is the stage for epidemiological reenactment and evaluation on geographic organizations. Epigrass is totally compiled in the Python language and utilize the NetworkX library to deal with the organizations. It gives pestilence models, like SEIR, SIR, SEIS, and SIS and a few varieties of these models
- **GEMF-sim:** GEMF-sim is the software apparatus that carries out the summed up plan of the outbreak spreading issue and the connected designing arrangement [27]. It is accessible in the well-known logical programming stages, like Python, C, MATLAB, and R. The models carried out cover the most widely recognized pestilence ones. It tends to be applied to break out measures with different hub contact and state layers; it permits clients to join relief procedures, for example, the appropriation of preventive practices and contact following the investigation of infection spreading
- **Nepidemix:** Nepidemix is the suite that customized to automatically portray reenactment of complex cycles on organizations. Nepidemix was created by individuals from the IMPACT-HIV bunch, and it is compiled in Python 2. The Nepidemix utilizes the module NetworkX to deal with the organization structure. At present, it gives three pestilence models: SIR, SIJR, and SIS. It automatizes the regular dissemination recreation steps permitting the software engineer to fabricate an organization as indicated by certain points of interest and to run in peak of it a bunch of pandemic cycles for a predetermined quantity of emphases. Besides, Nepidemix permits during execution to protect steady outcomes, like sickness predominance and state advances.
- **EoN:** EoN is the other widely utilized Python library committed to the execution of disseminating models. EoN is

intended to examine the breakout of SIR and SIS sicknesses in networks. It is made of two arrangements of algorithm: the principal set those arrangements with reenactment of scourges on networks (SIS and SIR) and the second that is intended to give arrangements of frameworks of conditions. Additionally, this bundle is based on top of NetworkX chart structures.

- **Epydemic:** Epydemic is also the other library developed for the executions of two scourge break out measures (SIR and SIS), reenacted over networks addressed utilizing NetworkX. It gives the essential recreation hardware to perform scourge reproductions under two distinctive reenactment systems: simultaneous reproduction in which time continues in discrete time spans and stochastic recreations.
- **ComplexNetworkSim:** ComplexNetworkSim is a Python package for the reenactment of specialists associated in the perplex network. The system is intended for clients having software engineering foundation; it can make a virtual complex organization with specialists that interface with one another. This task is not restricted to a static organization yet considers worldly organizations, where cycles can powerfully change the fundamental organization structure over the long haul. As of now, it gives two sorts of plague models: SIR and SIS.
- **Nxsim:** NXsim is a Python bundle for reenacting specialists associated by an organization utilizing NetworkX and SimPy in the Python 3.4. This research is a fork of the past ComplexNetworkSim package.
- **EpiModel:** Epimodel is quite possibly the most well-known package compiled in R. EpiModel allows the organization to construct, settle, and plot numerical models of irresistible infection. Right now, it gives usefulness to three classes of scourge models—speculative Individual interaction Models, speculative Network Designs and Deterministic Compartmental Models—and three sorts of irresistible illness can be reproduced upon them: SIS, SIR, SI. This bundle is based on top of iGraph network structures. EpiModel permits creating visual outlines for the execution of plague models; it gives plotting offices to show the methods and standard deviations across various recreations while shifting the underlying contamination status. It additionally incorporates an online visual application for reenacting.

- **RECON:** The RECON, R pandemic Confederation, gathers an assemblage of global specialists in irresistible sickness displaying, Public Health, and programming advancement to make the up-and-coming and next-generation apparatuses for infection episode investigation utilizing the R programming. The task incorporates the R bundle to figure, envision, and model infection episodes.
- **Sisspread:** Sisspread permits simulating the elements of a hypothetic irresistible infection inside a contact organization of associated individuals. It was compiled in C, and it carries out three traditional plans of infection development (SIS, SI, and SIR), which may assess the extension on various conveyance networks geographies (irregular homogeneous, without scale, little world) and, furthermore, on client gave networks.
- **GLEaMviz:** GLEaMviz is an openly accessible programming that recreates the break out of arising individual–individual irresistible infections on a world range [28]. The GLEaMviz structure is made out of three parts: the customer application, the intermediary middleware, and the recreation motor. The reenactments it characterizes consolidate true information on populaces and human versatility with intricate stochastic models of infection transmission to mimic sickness scattered on a worldwide scale. As yield, it gives a powerful guide and a few outlines portraying the geo-transient development of the infection. The recently recorded assets are intended to permit the last client to reenact plague models in organized settings following various reasonings. Be that as it may, because of the interdisciplinary idea of the particular issue handled, there are additionally a great deal of single model libraries expected to reproduce a particular illness or, alternately, broad reenactment instruments uncovering not many impromptu plague models
- **NetLogo:** NetLogo is a programmable designing environment for reproducing regular and social marvels. It was created by Uri Wilensky in 1999 [29] and has been in nonstop improvement from that point forward at the "Middle for Connected Learning and Computer-Based Modeling." It is especially appropriate for displaying complex frameworks that develop after some time, depicting them as specialist-based cycles. NetLogo empowers clients to operate a predetermined set of reproductions and distract with their

boundaries, investigating their practices under different conditions.

- **Framework Sciences:** System science or framework science is the online venture made by the "Organization of Systems Sciences, Innovation and Sustainability Research" developed at the Graz educational institution, which concentrates to plan an intelligent electronic course reading for frameworks sciences dependent on programming advantages for tablet PCs. In the illness break out segment suggested by this instrument, the client can pick an organization from a bunch of old style network representation (arbitrary, little world, sans complete, and scale organization) and afterward fix the boundary of the SIR model (the just one carried out up until now).
- **FRED:** The Framework for Reconstructing Epidemiological Dynamics is an open-access demonstrating framework created by the "Pitt Public Health Dynamics Laboratory" in a joint effort with the "Pittsburgh Supercomputing Center and the School of Computer Science" at Carnegie Mellon University. FRED upholds research on the elements of irresistible infection plagues and the interfacing impacts of moderation methodologies, viral development, and individual well-being conduct. The framework utilizes a specialist put together display based with respect to enumeration engineered populaces information that catch the segment and geographic appropriations of the populace. FRED plague models are, as of now, accessible for each state and country in the USA and for chosen global areas.
- **FluTE:** FluTE is a personal-dependent structure fit for recreating the break out of flu among significant metropolitan regions or the mainland USA [30]. It will reproduce a few intercession techniques, and these techniques will either alter the conveyance attributes of flu (e.g., inoculation) or alter the correspondent possibilities between people, such as social distancing. It is compiled in C++ or C.
- **Malaria Tool:** It is the UI to a joined mediation model for malarial fever, which was created by Imperial College London as a component of the Inoculation designing Initiative.
- **EpiFire:** Epifire is the rapid C++ implementation organizing interface for executing the spread of scourges on communication organizations.

- **Measles Virus:** It is a solicitation compiled in both Matlab and Python for the reenactment of the break out of the measles infection [31].

Contrasting various libraries is certainly not a simple errand. To be sure, the decision of hidden advancements, programming dialects, crowds just as conclusive points significantly shapes a setup of insightful apparatuses. In the accompanying, clients chose a subdivision of the recently presented systems and utilized a two-level examination enveloping both subjective and quantitative perspectives [18].

1.4 Execution of SNA in Terms of Real-Time Application: Implementation in Python

This section describes the application of the SNA using the Python libraries to a real-world application. For instance, let us consider the sentiment analysis of the social users in the COVID pandemic scenario or predicting and tracing the contiguous diseases. With the enhanced development of technology, the expected data can be attained by just typing the required keyword in the search engine. The number of sites of social networking is capable of providing more informative data that assist in the evaluation of SN. The data needed for the analysis are gathered through the application of data mining concept in social network sites. The creators of the social media platform, like Facebook, Reddit, Twitter, afford the users with Application Programming Interface (API) that assist in gathering the expected data from the website. Application Programming Interface acts as a medium of communication between the server and the client. It helps the creators to extract the data available in one location to the other with the provision of a function that assist in copying the data. The working principle of API differs from one programming language to the other. The data gathering, preprocessing, classification are the important stages in SNA, and it is depicted in Figure 1.3. Data gathering is the first step to execute any work in data mining. The process of data gathering is a flexible task, and it relies on the particular subject of user interest. Initially, the raw data are accumulated from the social network by requesting the data with a precise keyword.

After gathering the data from the social network, the data are preprocessed to execute the processes, like prediction or analysis. Based on the application, the collected data are processed with the preprocessing stages, and the data can be categorized and visualized. Nowadays, in Python, the classifiers implemented for an application is mainly any kind of the

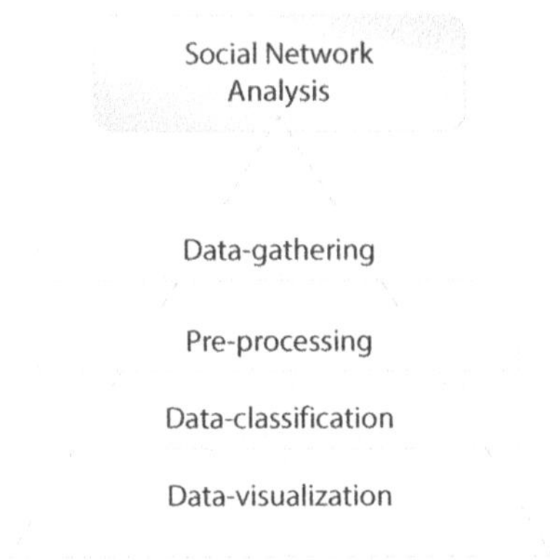

Figure 1.3 Flowchart of social network.

machine learning classifier that acts as a supervised machine learning approach. The classifier requires proper training using the labeled training data, without which the performance of the classifier cannot be analyzed. One of the commonly used statistical classifier is the Naïve Bayes classifier, which is generally used to classify the sentiments of people in COVID pandemic conditions. Such kind of classifiers generally utilizes the publicly available data (from the communal media data) in an efficient way to perform a prediction or analysis or classification problems.

1.5 Clarity Toward the Indices Employed in the Social Network Analysis

There are a number of metrics available for the SN analysis methods that measure the activity of the social users/nodes and ensure a better understanding of the analysis [32, 33]. Some of the metrics are discussed as follows:

1.5.1 Centrality

The evaluation of the constructional significance of a node present in an organization is executed using the metric called centrality. In other words, the preeminence of a node in an organization is deliberated using the centrality metrics. The highly influential people in the online social network can be identified using these metrics. A number of measures are used when evaluating the centrality metrics, such as Degree Centrality, Eigenvector Centrality,

PageRank measure, Between-ness Centrality, Closeness Centrality, and finally, the group Centrality [34].

1.5.2 Transitivity and Reciprocity

The linking characteristics of a network can be accessed using the transitivity and reciprocity metrics. The transitive nature between three edges can be analyzed using the transitivity metric in such a way to develop a triangle, and in the same way, the transitive nature of a node is analyzed using the reciprocity metrics.

1.5.3 Balance and Status

The consistency of the networks can be evaluated using the social balance and social status metrics. The social balance theory states that a friend relationship is consistent with the propagation of the transitivity among nodes as "the friend of my friend is my friend." Hence, the consistent triangles, depending on this strategy, are represented as balanced.

1.6 Conclusion

SN organization examination is the way toward researching social designs using organizations and chart hypothesis. It consolidates the assortment of strategies for examining the construction of interpersonal organizations just as speculations that target clarifying the hidden elements; furthermore, designs are seen in these constructions. It is an intrinsically interdisciplinary field, which initially rose up out of the fields of social brain research, insights, and chart hypothesis.

References

1. Mislove, A., Marcon, M., Gummadi, K.P., Druschel, P., Bhattacharjee, B., Measurement and analysis of online social networks, in: *Proceedings of the 7th ACM SIGCOMM conference on Internet measurement*, pp. 29–42, 2007.
2. Scott, J. and Carrington, P.J., The SAGE handbook of social network analysis. London: SAGE publications ltd, 2014.
3. Holme, P. and Saramäki, J., *Temporal networks, Physics reports*, vol. 519, pp. 97–125, 2012.
4. Lee, S., Rocha, L.E., Liljeros, F., Holme, P., Exploiting temporal network structures of human interaction to effectively immunize populations. *PloS One*, 7, 5, e36439, 2012.

5. Pennacchioli, D., Rossetti, G., Pappalardo, L., Pedreschi, D., Giannotti, F., Coscia, M., The three dimensions of social prominence, in: *Proceedings of International Conference on Social Informatics*, pp. 319–332, 2013.
6. Rossetti, G., Guidotti, R., Miliou, I., Pedreschi, D., Giannotti, F., A supervised approach for intra-/inter-community interaction prediction in dynamic social networks. *Soc Netw. Anal. Min.*, 6, 1, 1–20, 2016.
7. Camacho, D., Panizo-LLedot, Á., Bello-Orgaz, G., Gonzalez-Pardo, A., Cambria, E., The four dimensions of social network analysis: An overview of research methods, applications, and software tools. *Inform. Fusion*, 63, 88–120, 2020.
8. Akhtar, N., Social network analysis tools, in: *proceedings of Fourth International Conference on Communication Systems and Network Technologies*, pp. 388–392, 2014.
9. Mohr, I., The impact of social media on the fashion industry. *JABE*, 15, 2, 17–22, 2013.
10. Nash, J., Exploring how social media platforms influence fashion consumer decisions in the UK retail sector, *J. Fash. Mark. Manage*, 23, 1, 82–103, 2019. https://doi.org/10.1108/JFMM-01-2018-0012
11. Yu, Y. Moore, M. and Parillo-Chapman, L., Social media based, data-mining driven Social Network Analysis (SNA) of Printing Technologies in Fashion Industry, *International Textile and Apparel Association Annual Conference Proceedings*, 77, 1, 2020. https://doi.org/10.31274/itaa.11762
12. Kate, S., Wickremasinghe, D., Blanchet, K., Avan, B., Schellenberg, J., Use of social network analysis methods to study professional advice and performance among healthcare providers: a systematic review. *Syst. Rev.*, 6, 1, 1–23, 2017.
13. Wang, P., González, M.C., Menezes, R., Barabási, A.L., Understanding the spread of malicious mobile-phone programs and their damage potential. *Int. J. Inf. Secur.*, 12, 5, 383–392, 2013.
14. Burt, R.S., Social contagion and innovation: Cohesion versus structural equivalence. *Am. J. Sociol.*, 92, 6, 1287–1335, 1987.
15. Milli, L., Rossetti, G., Pedreschi, D., Giannotti, F., Information diffusion in complex networks: The active/passive conundrum, in: *Proceedings of International Conference on Complex Networks and their Applications*, pp. 305–313, 2017.
16. Sîrbu, A., Loreto, V., Servedio, V.D., Tria, F., Opinion dynamics: models, extensions and external effects, in: *Participatory Sensing, Opinions and Collective Awareness*, pp. 363–401, 2017.
17. Sîrbu, A., Loreto, V., Servedio, V.D., Tria, F., Opinion dynamics with disagreement and modulated information. *J. Stat. Phys.*, 151, 1, 218–237, 2013.
18. Rossetti, G., Milli, L., Rinzivillo, S., Sîrbu, A., Pedreschi, D., Giannotti, F., NDlib: a Python library to model and analyze diffusion processes over complex networks. *Int. J. Data Sci. Anal.*, 5, 1, 61–79, 2018.

19. Staudt, C.L., Sazonovs, A., Meyerhenke, H., NetworKit: A tool suite for large-scale complex network analysis. *Netw. Sci.*, 4, 4, 508–530, 2016.
20. Hogan, B., Visualizing and interpreting Facebook networks, in: *Analyzing Social Media Networks with NodeXL (2010)*, Morgan Kaufmann, Massachusetts.
21. Gunawan, T.S., Abdullah, N.A.J., Kartiwi, M., Ihsanto, E., Social network analysis using python data mining, in: *Proceedings of 8th International Conference on Cyber and IT Service Management (CITSM)*, pp. 1–6, 2020.
22. Viard, T., Latapy, M., Magnien, C., Computing maximal cliques in link streams. *Theor. Comput. Sci.*, 609, 245–252, 2016.
23. Housley, W., Procter, R., Edwards, A., Burnap, P., Williams, M., Sloan, L., Rana, O., Morgan, J., Voss, A., Greenhill, A., Big and broad social data and the sociological imagination: A collaborative response. *Big Data Soc.*, 1, 2, 2053951714545135, 2014.
24. Casteigts, A., Flocchini, P., Quattrociocchi, W., Santoro, N., Time-varying graphs and dynamic networks, *Int. J. Parallel Emergent Distrib. Syst.*, 27, 5, 387–408, 2012.
25. Ackland, R. and Zhu, J.J., Social network analysis, in: *Innovations in Digital Research Methods*, pp. 221–244, 2015.
26. Goldenberg, D., *Social Network Analysis: From Graph Theory to Applications with Python*. PyCon'19. arXiv preprint arXiv: 2102.10014, 2021
27. Sahneh, F.D., Vajdi, A., Shakeri, H., Fan, F., and Scoglio, C., GEMFsim: A stochastic simulator for the generalized epidemic modeling framework. *J. Comput. Sci.*, 22, 36–44, 2017.
28. Van den Broeck, W., Gioannini, C., Gonçalves, B., Quaggiotto, M., Colizza, V., Vespignani, A., The GLEaMviz computational tool, a publicly available software to explore realistic epidemic spreading scenarios at the global scale. *BMC Infect. Dis.*, 11, 1, 1–14, 2011.
29. Wilensky, U. and Tisue, S., Netlogo: A simple environment for modeling complexity, in: *Proceedings of International conference on complex systems*, vol. 21, pp. 16–21, 2004.
30. Chao, D.L., Halloran, M.E., Obenchain, V.J., Longini Jr., I.M., FluTE, a publicly available stochastic influenza epidemic simulation model. *PloS Comput. Biol.*, 6, 1, e1000656, 2010.
31. Word, D.P., Abbott, G.H., Cummings, D., Laird, C.D., Estimating seasonal drivers in childhood infectious diseases with continuous time and discrete-time models. *Proceedings of the American Control Conference*, pp. 5137–5142, 20102010.
32. Zafarani, R., Abbasi, M., & Liu, H., Social media mining: An introduction. Cambridge: Cambridge University Press, 2014.
33. Sahu, B.P., Gouse, M., Pattnaik, C.R., Mohanty, S.N., MMFA-SVM: New bio-marker gene discovery algorithms for cancer gene expression. *Materials Today: Proceedings*, 2021, https://doi.org/10.1016/j.matpr.2020.11.617.
34. Arulkumar, N., Galety, M.G., Manimaran, A., CPAODV: Classifying and assigning 3 level preference to the nodes in VANET using AODV based

CBAODV algorithm, in: *Intelligent Computing Paradigm and Cutting-edge Technologies. ICICCT 2019. Learning and Analytics in Intelligent Systems*, L. Jain, S.L. Peng, B. Alhadidi, S. Pal (Eds.), Springer, Cham, 2020, vol. 9, https://doi.org/10.1007/978-3-030-38501-9_42.

2

Introduction To Python for Social Network Analysis

Agathiya Raja[1]*, Gavaskar Kanagaraj[1] and Mohammad Gouse Galety[2]

[1]Computer Science, Technical University of Clausthal, Clausthal-Zellerfeld, Germany
[2]Department of Information Technology, Catholic University in Erbil, Erbil, Iraq

Abstract
A social network is an architecture that consists of the communication among actors, which holds further information about their details and relationship with one another. They are interconnected in the form of edges (or link) and nodes (or vertices). Every social network has its purposes like education, business, consulting, and so on. Social networking platforms play an ever-increasing vital role in almost every field of daily life, including past predictions to future technologies. The intense use of social networking platforms provides a good understanding overview of the community and social behavior. However, well-known projections and conclusions based on analyzing social networking platforms tend to be inexact. A study or analysis on the social network is helpful in many ways (e.g., to find the criminal). Using network-level analysis, one could isolate an objective component/node in a network. One could identify the core, density. One could compute the shortest path, reciprocity, and even homophily. There are incompatible properties among the networks and the network resemblance or connection between multiple networks. Analyzing and visualizing the network using Python offer good insights about the networks to end-users. A high-level programming language provides significant advantages for the end-users and tender vast library packages for integration. Python is an uncomplicated interpreter language, and it is fast to prototype. The language is proposed with several algorithms, which are used to analyze the complex graph. It is incorporated with many packages and libraries, each possessed to perform the desired methodology. The chapter explains the installation and working environment of Python.

***Keywords*:** Python, social network analysis, Network-X, graph, nevaal

**Corresponding author*: agathiya.raja@tu-clausthal.de

Mohammad Gouse Galety, Chiai Al Atroshi, Bunil Kumar Balabantaray and Sachi Nandan Mohanty (eds.) Social Network Analysis: Theory and Applications, (19–36) © 2022 Scrivener Publishing LLC

2.1 Introduction

A *social network* is an architecture that consists of the communication among actors, which holds further information about their details and relationship with one another. They are interconnected in the form of edges (or links) and nodes (or vertices). Every social network has its purposes, like education, business, consulting, and so on. Social networking platforms play an ever-increasing vital role in almost every field of daily life, including past predictions to future technologies. The intense use of social networking platforms provides a good understanding overview of the community and social behavior. However, well-known projections and conclusions based on analyzing social networking platforms tend to be inexact.

A study or analysis on the social network is helpful in many ways (e.g., to find the criminal). Using network-level analysis, one could isolate an objective component/node in a network. One could identify the core, density. One could compute the shortest path, reciprocity, and even homophily. There are incompatible properties among the networks and the network resemblance or connection between multiple networks.

Analyzing and visualizing the network using Python offers good insights about the networks to end-users. A high-level programming language provides significant advantages for the end-users and tender vast library packages for integration. Python is an uncomplicated interpreter language, and it is fast to prototype. The language is proposed with several algorithms which are used to analyze the complex graph. It is incorporated with many packages and libraries, each possessed to perform the desired methodology. This chapter explains the installation and working environment of Python.

NetworkX is one of the most efficient software packages and an open-source tool in Python. It is mainly used to analyze the complex graph database by manipulating the larger data sets. The chapter explains more about the importance of using Python with desired examples. The installation setup and working environment have been clearly explained in this chapter for better understanding. Although NetworkX is not ideal for large-scale problems with fast-processing requirements, it is an excellent option for real-world network analysis, standard graph algorithms. It is optimum for denoting networks of different types, like classic graphs, random graphs, and artificial networks. Thus, it takes advantage of Python's ability to import data from external sources. The package NetworkX consists of many functions of graph generators and facilities to manipulate (read and write) the graphs in so many formats, such as .edgelist, .adjlist, .gml, .graphml, .pajek, and so on.

In the last section, we will provide glimpses of NetworkKit. NetworKit is also a Python module, and it can be used as an alternative for NetworkX. In NetworkKit, performance-aware algorithms are written in C++ (often using OpenMP for shared-memory parallelism) and exposed to Python via the Cython toolchain.

2.2 SNA and Graph Representation

A network, in general, is a set of objects/nodes with interconnections. Now thinking about why we want to study/analyze networks because networks are everywhere, for examples, social networks, friendship networks, email communication networks, Networks of relatives. Transportation and Mobility Networks [1], Information Networks, Internet, websites relations, Biological networks, protein interactions, financial networks, and so on. Network analysis helps in understanding complex phenomena.

2.2.1 The Common Representation of Graphs

a. Undirected:
   The edges do not have any directions.
   Directed Networks:
   The edges have directions.

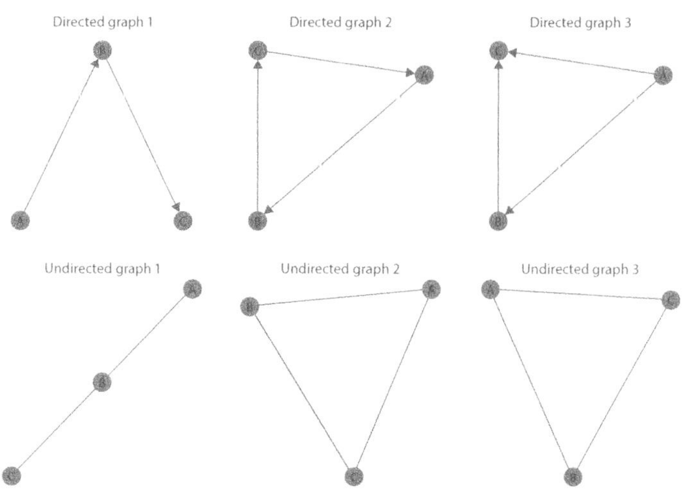

Figure 2.1 Comparison directed and undirected graph.

b. Simple:
   The graph has only one link type.

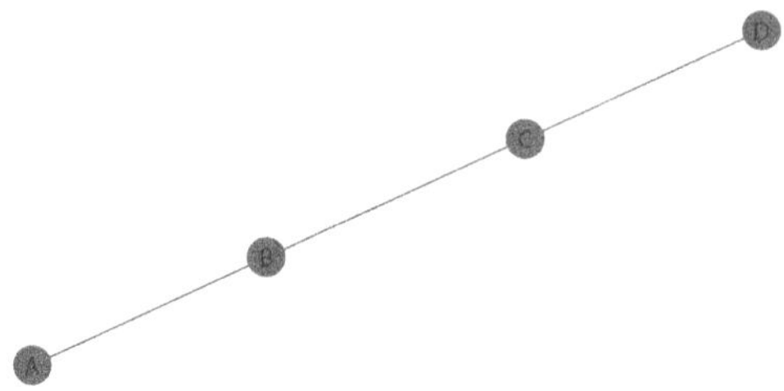

Figure 2.2 Simple graph.

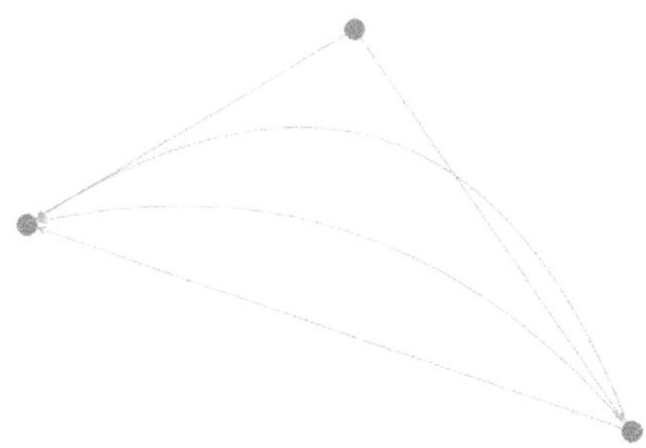

Figure 2.3 Multigraph.

   Multigraph:
   The graph can have more than one same link type.

c. Unweighted:
   The edges in the graph do not contain weight.
   Weighted:
   The edge in the graph contains value (numerical), which is known as weight.

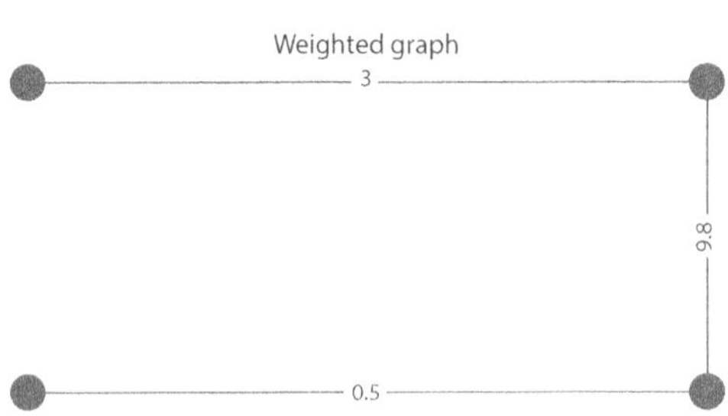

Figure 2.4 Weighted graph.

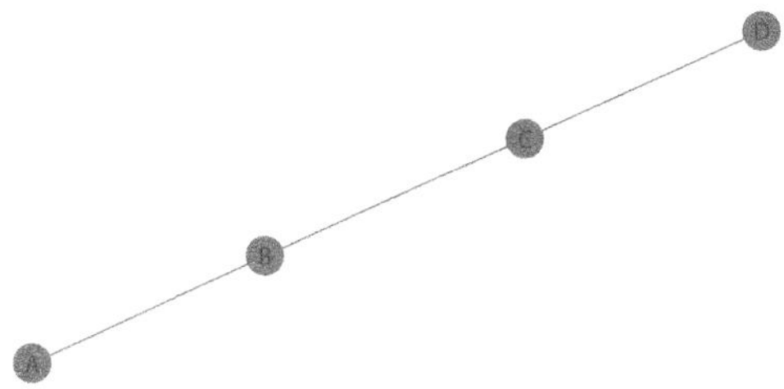

Figure 2.5 Unweighted graph.

Other important graphs:

i. Regular graph,
ii. Complete graph,
iii. Path,
iv. Cycle,
v. Bipartitie graph,
vi. Euler graph,
vii. Hamilton graph,
viii. Planner,
ix. Tree and forest, and so on.

2.2.2 Important Terms to Remember in Graph Representation

a. Centrality measures

Centrality measures are a significant indicator used in network analysis. There are different types of centrality measures. Some prominent measures are given as follows:

- Betweenness centrality
  Assuming the important nodes connect other nodes. The betweenness centrality is defined as the cumulative sum of ratios of the paths between two nodes through a node to the total number of shortest paths available between those nodes.
- Closeness centrality
  Assuming in a connected graph, closeness centrality is a measure of centrality in the given network. The node is closer to all nodes if it is more central.

- Degree centrality
  Assuming the networks where all nodes are connected and one or more than one nodes have predominant connections in comparison with other neighbouring nodes For instance, in an undirected graph, the degree centrality is defined by the number of connections attached to each node.
- Eigenvector centrality
  Assuming the networks where all nodes are connected and one or more than one nodes have predominant connections in comparison with other neighboring nodes. Eigenvector centrality is an algorithm that measures the influence or connectivity of nodes [2]. Relationships to high-scoring nodes contribute more to the score of a node than connections to low-scoring nodes. A high score means that a node is connected to other nodes that have high scores.
- PageRank centrality
  Assuming the networks where all nodes are connected and one or more than one node have predominant connections in comparison with other neighboring nodes. For instance, nodes relate to links representing appropriate weights and weights are updated when the node centrality/significance changes in the directed network [3].

b. Geodesic distance
c. Networks

- Distributed
- Centralized
- Decentralized

2.3 Tools To Analyze Network

There are some traditional and basic tools, which are still helpful to analyze the small network. The following explanations will provide the advantages and limitations of two traditional tools.

2.3.1 MS Excel

MS Excel is one of the basic software packages from Microsoft, which is popular among the users. Along with the mathematical operations, it can also be used to analyze social network [4]. The packages like NodeXL and. NetMap are used exclusively for network analysis in MS Excel.

NodeXL

NodeXL is intended to be used by user with less programming experience. Various graph formats, like UCINet.dl, edgelists, and so on, have been supported by NodeXL.

Some of the activities that can be done with this package are as follows:

a. data importing,
b. data representation,
c. graph analysis,
d. graph visualization.

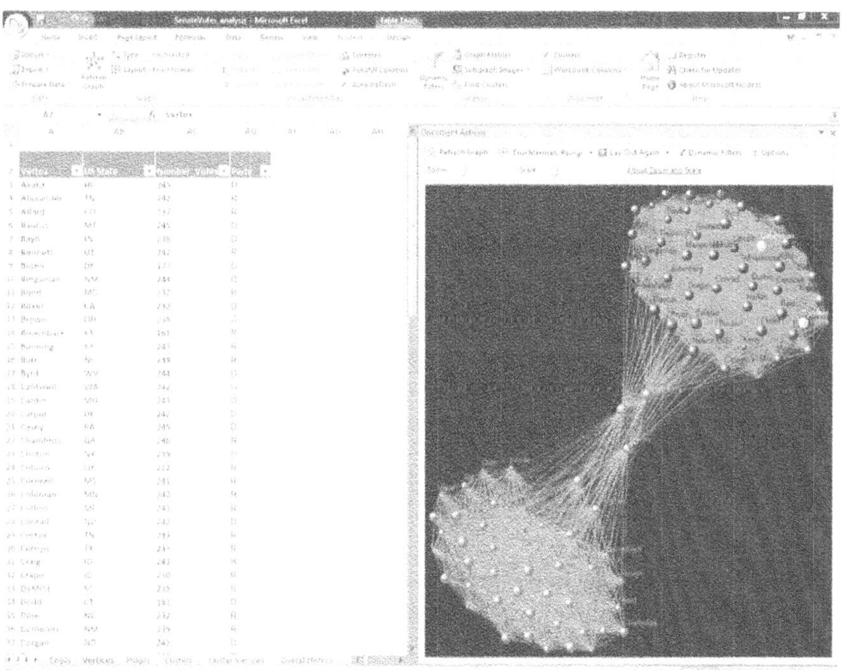

Figure 2.6 NodeXL network overview discovery and exploration in excel [5].

NetMap

It is another social networking tool for Excel, which is fast and user friendly. It is highly used to map the network visually [6]. The main motivation of the tool is to find the *linkage among the partners*. It is calculated based on the linkage towers. If the linkage tower is higher, then the partners are highly influential and vice versa.

2.3.2 UCINET

It is an extensive network analysis package, which can be run in Windows. However, other machines, like Linux or Mac, can also use it under VMWare or Parallel or BootCamp. It includes a visualization tool called ***NetDraw***, which shows the picturized output of the network.

Summary

The overall drawback of this traditional tools are as follows [7]:

1. It is maily used only in Windows,
2. Requires more RAM for better performance,
3. Can be good for small data set,
4. Too slow to run more nodes.

2.4 Importance of Analysis

Analyzing the social network is significant to monitor the activities among the groups. In recent days, SNA has been used to find a strategic application to build better team. Cultural issues and influences within a group can be understand which will help to structure the world better place [8]. Deep understanding of biological systems, change in natural phenomena, target terrorism, and fake ID detection can be achieved with the help of Social Network Analysis (SNA).

2.5 Scope of Python in SNA

Python is trending, as well as the most demanding knowledge in recent years. Python is also the most wanted language. The community is growing quite fast. In the past decades, social networks were analyzed using frameworks like c. As we know, scope and applications of social networks are increasing drastically [9]. It is not practically feasible that domain experts from different fields also need expertise in programming to implement their ideas of network analysis; however, learning python as a tool to empower their ideas supporting visions is realizable [10].

This section introduces the prominent syntax and syntax styles of python, as well as different library packages and its significance [11].

2.5.1 Comparison of Python With Traditional Tools

1. Python is free open source, whereas MS Excel is a paid package.
2. Python is easy for a complex equation and a huge data set, and Excel is good only for a small data set.
3. Since python is an open source, anyone can audit or replicate a work that is not possible in Excel.
4. Finding errors and debugging it is a lot easier in Python than Excel.
5. Excel is way simpler to use than Python, i.e., the user does not need any programming knowledge.
6. Repetitive tasks can be easily done by *automation*, which is not possible in Excel.
7. Python provides in-depth visualizations, whereas Excel has basic graphs [12].

2.6 Installation

Python installation consumes a bit more time because it should be properly downloaded in the right environment with all the necessary packages [13]. The standard version of python can be installed from the following link [https://www.python.org/downloads/].

Different versions of Python with respect to the type of OS (Windows, Mac, Linux) can be found under this link.

Some important package for SNA is pandas, matplotlib, and NetworkX. All these packages can be installed via pip installation.

- pip install pandas
- pip install network
- pip install matplotlib

NetworkX is an important library used to analyze social network in Python [14]. The package is mainly created to analyze the functions of complex graph structure. It is a free package under BSD license.

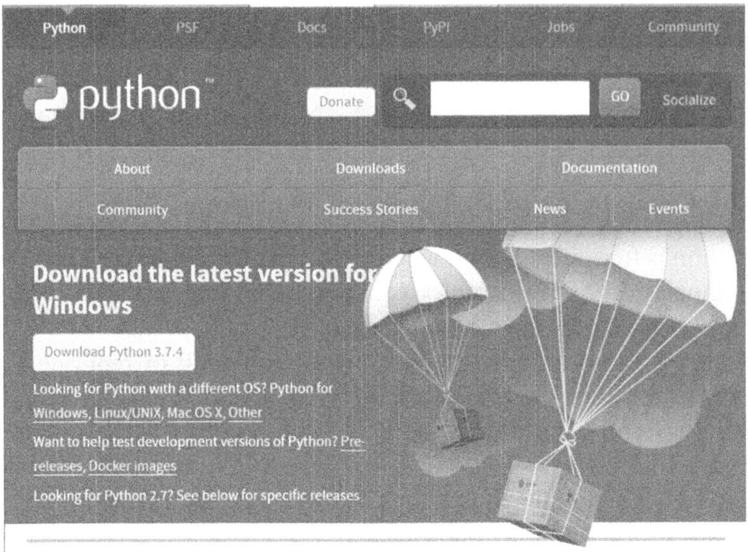

Figure 2.7 Python official documentation.

2.6.1 Good Practices

1. It is always advised to install virtual environments like *Anaconda environment. Miniconda* can be used instead of anaconda if the computer has less than 5 Gb ram [15]. You can download the standard version of Anaconda here [https://docs.anaconda.com/anaconda/install/].
2. Choosing editors, such as VS code or pycharm or IntellIj or Jupyter Notebooks, and so on, comes along with the Anaconda environment.
3. Proceed with open-source version at the beginning.
   Use Anaconda Navigator→ interactive Visual mode
   Or Prompt Terminal Mode:

   - Creating new environments in Anaconda: conda create—name myenv
   - Replace myenv with the environment name.
   - Activate Environment: conda activate myenv
   - Installing packages: conda install [packagename]

The more useful resources and explanations on working with *conda environment* can be found in their official documentation.

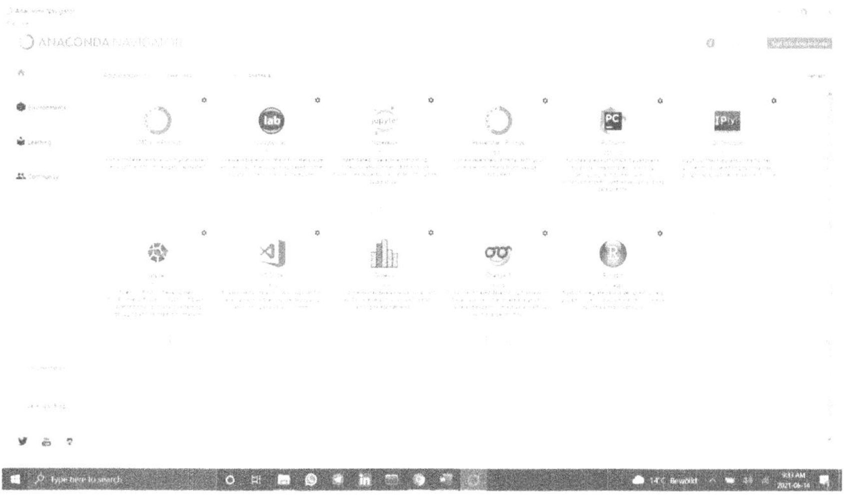

Figure 2.8 Anaconda navigator.

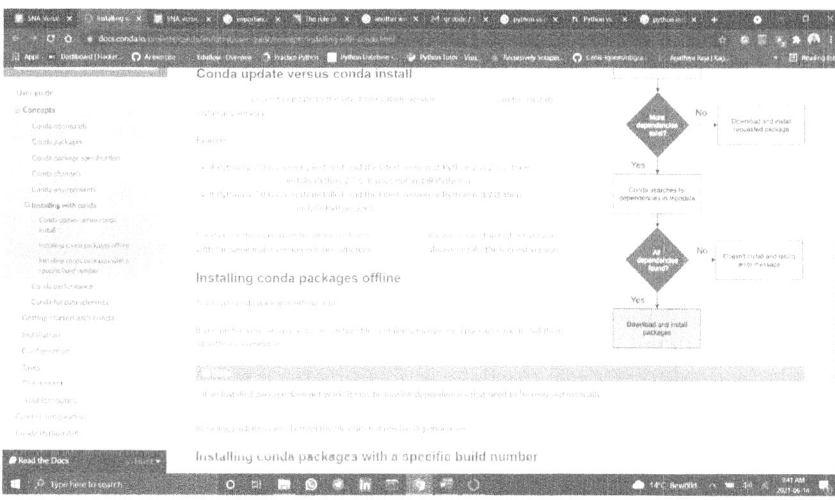

Figure 2.9 Conda environment installation.

2.7 Use Case

Some interesting case studies based on SNA are Facebook friends' group and terrorist activities [16]. The case study has been worked in python with Jupyter notebook. You can download and explore the data set to get more insight under the following link.

Scan the QR code and follow the Github link to access the worksheets.

Figure 2.10 QR code for workbooks and source codes.

2.7.1 Facebook Case Study

The first important steps in analyzing any kind of data set in python is *importing libraries.* The data to be analyzed can be scrapped directly from the respective site or it can be accessed from the API provided by the website [17]. Choosing the data mainly depends on the need, i.e., why do we need to analyze the data? What is the purpose? What kind of problem are we solving? [18]

Step 1: Import libraries

Each library has their built-in function, which makes Python easy to code.

```
In [3]:  import pandas as pd
         import networkx as nx
         import matplotlib.pyplot as plt
         %matplotlib inline
         import warnings; warnings.simplefilter('ignore')
```

Figure 2.11 Code blocks for importing libraries.

Step 2: Read data

Pandas is used to retrieve the data and can be used to explore a huge data set conveniently.

```
In [2]:  df = pd.read_csv('facebook_combined.txt')
         #df
         df.info()
         df.tail()

         <class 'pandas.core.frame.DataFrame'>
         RangeIndex: 88233 entries, 0 to 88232
         Data columns (total 1 columns):
          #   Column  Non-Null Count  Dtype
         ---  ------  --------------  -----
          0   0 1     88233 non-null  object
         dtypes: object(1)
         memory usage: 689.4+ KB

                    0 1
         88228  4026 4030
         88229  4027 4031
         88230  4027 4032
         88231  4027 4038
         88232  4031 4038
```

Figure 2.12 Code block for reading data.

Step 3: Data cleaning

Data cleaning means removing/ cleaning the noise (NaN, Missing data) [19]. Data quality will have more impact in the model so using the data with less noise is recommended for better results. Missing values can be altered by generating the mean, median value and so on [20–22]. It completely depends upon the type of data.

Step 4: Read input

read_edgelist is a built-in function in NetworkX library. More details about it can be found in the documentation website. [23]

```
In [11]:  G_fb = nx.read_edgelist("facebook_combined.txt", create_using = nx.Graph(), nodetype=int)
          print(nx.info(G_fb))

          Name:
          Type: Graph
          Number of nodes: 4039
          Number of edges: 88234
          Average degree:  43.6910
```

Figure 2.13 Code block for reading edge list.

Step 5: Visualizing the network

```
In [16]:  plt.figure(figsize=(20,20))
          nx.draw_networkx(G_fb);
```

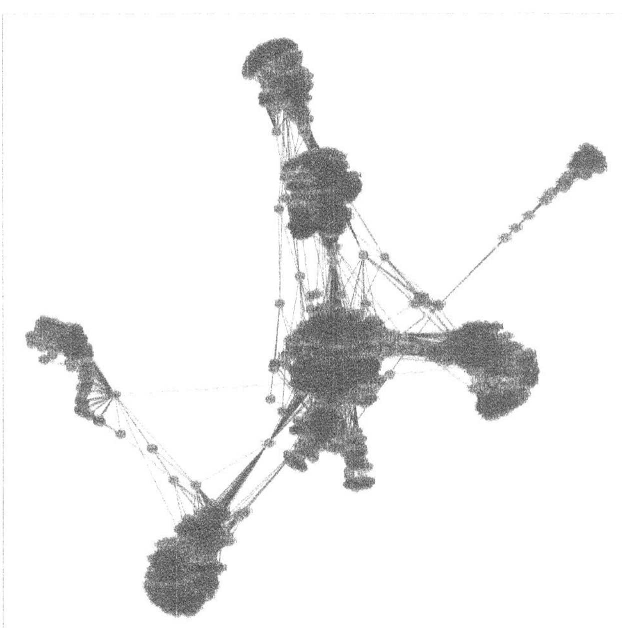

Figure 2.14 Visualization of Facebook users.

Step 6: Centrality measures

```
In [17]:  pos = nx.spring_layout(G_fb)
          betCent = nx.betweenness_centrality(G_fb, normalized=True, endpoints=True)
          node_color = [20000.0 * G_fb.degree(v) for v in G_fb]
          node_size =  [v * 10000 for v in betCent.values()]
          plt.figure(figsize=(20,20))
          nx.draw_networkx(G_fb, pos=pos, with_labels=False,
                           node_color=node_color,
                           node_size=node_size )
          plt.axis('off');
```

Figure 2.15 Code block for centrality measures.

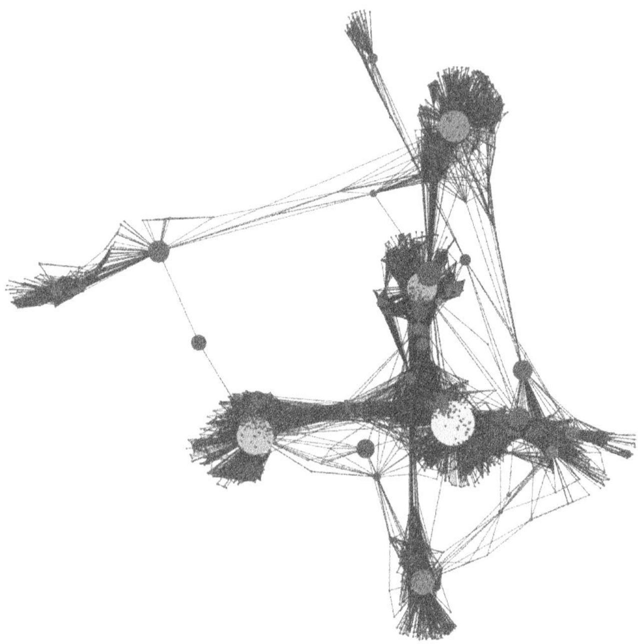

Figure 2.16 Visualization of centrality measures on Facebook users.

2.8 Real-Time Product From SNA

One of the innovative and fancy real-time products out of network analysis is nevaal maps, which is created by nevaal AG, a German company focused mainly on network analysis for business.

Company Vision:

The motive of the company is to "create a front-line solution to visualize information from our social circles."

2.8.1 Nevaal Maps

It is the SaaS application used in business network analytics. It connects the network (group of people) in the business network together to track them, getting in touch and to make better decision. The capability of it to handle the complex data makes it easier for any start-up to keep their organization in a structured manner.

The three important features about nevaal maps, which makes it more efficient, are as follows: *scalable, secure, and customizable.* The central mechanism can be adjusted according to individual customer need.

Usage

Visualizing the complex network data helps in

- Screening process and investment decisions.
- Enabling the internal/external process of data.
- Providing interactive and insightful view of the business data.

Significancy

The product is not only focusing on visualizing the network connection but also aids in manifesting communication processes, which is outcome focused.

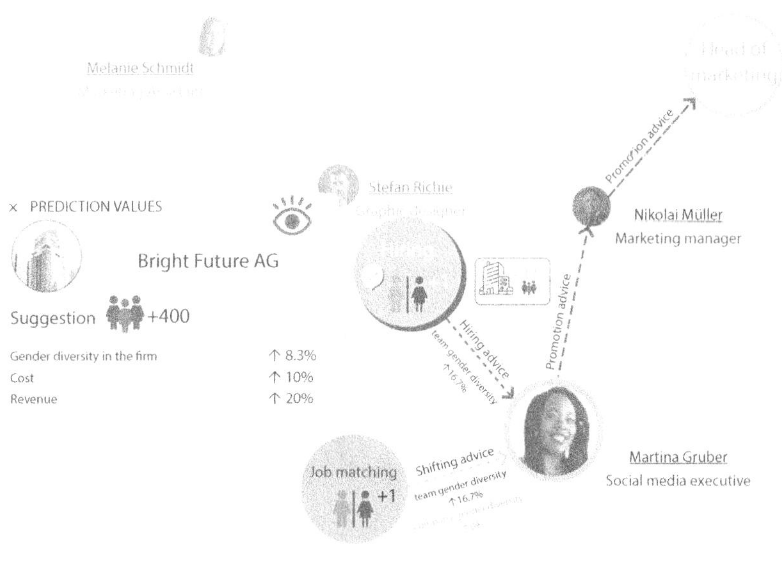

Figure 2.17 Visualization of graph database used in business.

Social network analysis helps us in every domain, such as fake ID detection, terrorist activities, marketing, social media, and so on.

References

1. Mona, E., Hari, R.M., Somya, V., Sivakumari, M.S., Alumni Social Networking Site. *Int. J. Sci. Res. Comput. Sci. Eng. Inf. Technol.*, 7, 467–472, 2021.
2. Otte, E. and Rousseau, R., Social network analysis: a powerful strategy, also for the information sciences. *J. Inf. Sci.*, *28*, 441–453, 2002.
3. Scott, J., Social Network Analysis. *Sociology*, *22*, 109–127, 1988.
4. McGloin, J. and Kirk, D., An Overview of Social Network Analysis. *J. Crim. Justice Educ.*, *21*, 169–181, 2010.
5. https://www.microsoft.com/en-us/research/project/nodexl-network-overview-discovery-and-exploration-in-excel/
6. Burcher, M., *Social Network Analysis and the Characteristics of Criminal Networks*, Australia, 2020.
7. Palus, S. and Kazienko, P., Social Network Analysis in Corporate Management, in: *MISSI*, 2010.
8. Wasserman, S. and Faust, K., *Social Network Analysis: Methods and Applications*, 1994.
9. Robins, G., A tutorial on methods for the modeling and analysis of social network data. *J. Math. Psychol.*, *57*, 261–274, 2013.
10. Gunawan, T.S., Abdullah, N.A., Kartiwi, M., Ihsanto, E., Social Network Analysis using Python Data Mining. *2020 8th International Conference on Cyber and IT Service Management (CITSM)*, pp. 1–6, 2020.
11. Haythornthwaite, C., Social network analysis: An approach and technique for the study of information exchange☆. *Libr. Inf. Sci. Res.*, *18*, 323–342, 1996.
12. Goldenberg, D., *Social Network Analysis: From Graph Theory to Applications with Python*, 2021, ArXiv, abs/2102.10014.
13. Hagberg, A., Schult, D., Swart, P., *Exploring Network Structure, Dynamics, and Function using NetworkX*, 2008.
14. Staudt, C., Sazonovs, A., Meyerhenke, H., NetworKit: A tool suite for large-scale complex network analysis. *Netw. Sci.*, *4*, 508–530, 2016.
15. Aslak, U. and Maier, B., Netwulf: Interactive visualization of networks in Python. *J. Open Source Software*, *4*, 1425, 2019.
16. Brandes, U., A faster algorithm for betweenness centrality. *J. Math. Sociol.*, *25*, 163–177, 2001.
17. Bader, D.A., Kintali, S., Madduri, K., Mihail, M., Approximating Betweenness Centrality, in: *WAW*, 2007.
18. Green, O., McColl, R., Bader, D.A., A Fast Algorithm for Streaming Betweenness Centrality. *2012 International Conference on Privacy, Security,*

Risk and Trust and 2012 International Confernece on Social Computing, pp. 11–20, 2012.
19. López-Acosta, A., García-Hernández, A., Vázquez-Reyes, S., Mauricio-González, A., A Metadata Application Profile to Structure a Scientific Database for Social Network Analysis (SNA). *2020 8th International Conference in Software Engineering Research and Innovation (CONISOFT)*, pp. 208–215, 2020.
20. Romero-Moreno, L., Methodology with Python Technology and Social Network Analysis Tools to Analyze the Work of Students Collaborating in Facebook Groups. *2019 14th Iberian Conference on Information Systems and Technologies (CISTI)*, pp. 1–6, 2019.
21. Krishna, R.P.M., Mohan, A., Srinivasa, K., Practical Social Network Analysis with Python, in: *Computer Communications and Networks*, 2018.
22. Liu, X., Sun, T., Bu, F., Qin, H., The Analysis on the Role of Social Network in the Field of Anti-Terrorism Take the "East Turkistan" Organization as an Example. *2020 5th International Conference on Mechanical, Control and Computer Engineering (ICMCCE)*, pp. 2282–2285, 2020.
23. Siddalingappa, K., Debabrata, S., Mohammad, G.G., Shivamurthaiah, M., A Hybridization Approach based Semantic Approach to the Software Engineering. *Test Eng. Manage.*, 83, March–April, 5441–5447, 2020.

3

Handling Real-World Network Data Sets

Arman Abouali Galehdari[1]*, Behnaz Moradi[1] and Mohammad Gouse Galety[2]

[1]Department of Informatik, Technical Universität Clausthal, Clausthal-Zellerfeld, Germany
[2]Department of Information Technology, College of Engineering Technology, Catholic University in Erbil, Kurdistan Region, Iraq

Abstract
There are certain ideas that must be clarified before anything else to have a better grasp of the topic. What are network data sets, and how do they differ from one another? What is real-world network data sets and how do they work and what exactly does they depend on? Also, other topics, such as network graphs, descriptions of networks, how they appear, and how they will be formed, are also discussed in detail. The introduction of the Scale-Free Network as one of the primary scenarios for modeling and analyzing the performance of the network was a significant step in the right direction and a description of small-world phenomenon to provide an approximate data from the massive network, such as social media. As a result, it has come to a conclusion on the possibilities and solutions for dealing with this kind of network and scenario, which basically are in massive size.

Keywords: Real-world network data sets, graph, scale-free network, small-world phenomenon, node, social media, network data sets, network model

3.1 Introduction

In today's modern world, large data sets are frequently created and applied for a variety of purposes, whether they are in the form of gigabytes of data contained in a single file or hundreds of files, each containing a small amount of data, with both of these approaches encountering difficulties and issues that must be addressed and resolved in order to provide the

**Corresponding author*: arman.abouali.galehdari@tu-clausthal.de

Mohammad Gouse Galety, Chiai Al Atroshi, Bunil Kumar Balabantaray and Sachi Nandan Mohanty (eds.) Social Network Analysis: Theory and Applications, (37–50)

required performance. The massive size of data, which is typically unsuitable for processing and management by a single machine and cannot be contained in a single file.

Real-world networks are large in size and intricate in structure, which complicates data handling and management. As examples of this type of data, millions or even billions of newly created user accounts on popular social media platforms, such as Facebook, Instagram, and Twitter, among others, could provide a reasonable estimate of the size of this subject's data, where the importance of these data sets are to analyze or improve their activity and troubleshoot problems.

These data would be useful if they could be analyzed to gain better information and come up with better ideas to improve the structure, such as when Facebook is interested in finding out about the number of new users who are active and those who are less active. This would be a working model that needs to be analyzed and run to produce the desired results.

A network consists of nodes or elements and the edges that connect them; in other words, the structural relationships between these nodes define and create the network's structure. Indeed, each of these created user accounts will be represented by a node, and the connections between them will define the network and its behavior.

Real-world network data sets are created to simulate the structure of the real world, which we encounter on a daily basis. Although mathematical equations and network models are used to analyze and handle this type of data, the complexity of this system does not allow for precise numerical data; however, approximate results are useful when sampling a network of this size.

Alternatively, dealing with big amounts of data may result in out-of-memory issues, the creation of needless blank spots on the screen, and sluggish functionality. The practice of performing long-term monitoring from permanent gauges and high-frequency readings in the field is now becoming more common, resulting in data sets that are substantially bigger than those necessary for pressure transient analysis [1].

3.2 Aspects of the Network

The term "connection" has been the subject of increasing public fascination in contemporary world society over the last decade, with the concept of a network at its heart. A pattern of interconnections between a collection of objects is a definition of this concept. To put it another way, a network is any collection of things in which certain pairs of these items are linked to another via links. Numerous different types of relationships or connections can be used to define links, depending on the context [2].

Assume that each email address is a system object; thus, a massive number of accounts exist in this network assumption. These accounts, on the other hand, are unable to provide information; they are analogous to a collection of unused dots in the environment. This inactive information can be generalized by adding links or, in the other words, connections to the dots that indicate the relationship between the objects. In this case, when an email is sent to one or more other email addresses, a network is created because we have objects and their relationships. More precisely, this system's objects are referred to as nodes or elements of the network, whereas the lines are referred to as edges. In other words, when two or more nodes discover a relationship between them, edges are formed. Indeed, a network is formed when nodes and defined relationships between them exist to provide a description of the network.

The purpose of Figure 3.1 is to demonstrate how a network in small size looks like, as well as to provide a better vision and understanding of a large-scale network. The small network depicted in Figure 3.1 represents the friendship connection between 34 members of a karate club. Each member of this club is represented by a numerical circle, and the friendship that exists between them is demonstrated by the connection.

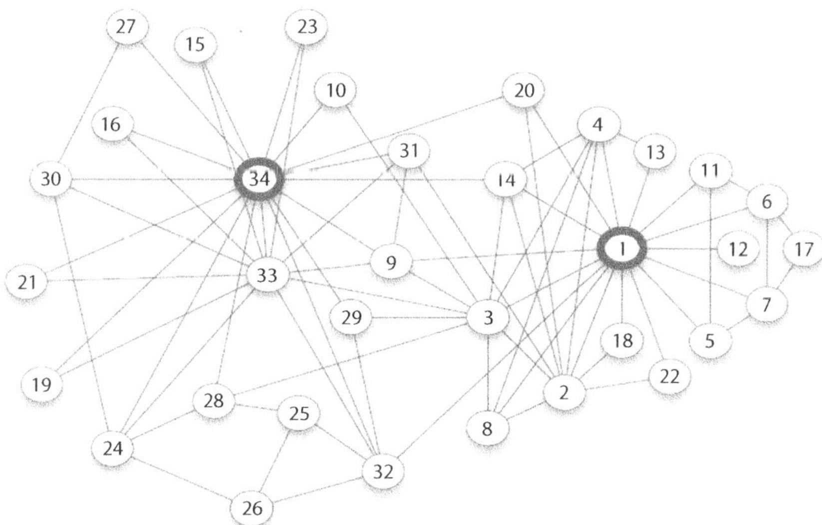

Figure 3.1 The presentation of a small-scale network based on the interpersonal relationships of members of a karate club [2].

For example, number 20 is acquainted with numbers 1, 2, and 34, which means that number 20 is acquainted with three other members of this club.

When the network grows to a larger size, the appearance, as well as the situation, will be changed as shown in Figures 3.2 and 3.3. The graph below presents e-mail exchange among a company employee and the links between web blogs. For each scenario, the nodes and the edges will be different from other situations, hence, nodes are presenting the e-mail address of the company staff and the connection between the nodes are the exchange of e-mail between those two addresses. The size of the network regarding their behavior may have different shape and interactions.

When analyzing a network, one of the most important aspects to consider is the network's structure, which can vary depending on the situation. Defining precise for a network is as difficult as studying it, and for the analyzer, understanding the structure and behavior of the network and the interaction between the nodes and their connections is important.

In network science and network theory, dynamic network analysis (DNA) is a newly emerging scientific field that brings together traditional social network analysis (SNA), link analysis (LA), social simulation, and multiagent systems (MAS) with traditional link analysis (LA). It is important to note that there are two aspects to this field.

The first is a statistical analysis of DNA data. The second is the application of simulation to address issues relating to the dynamics of network systems. DNA networks differ from traditional social networks in that

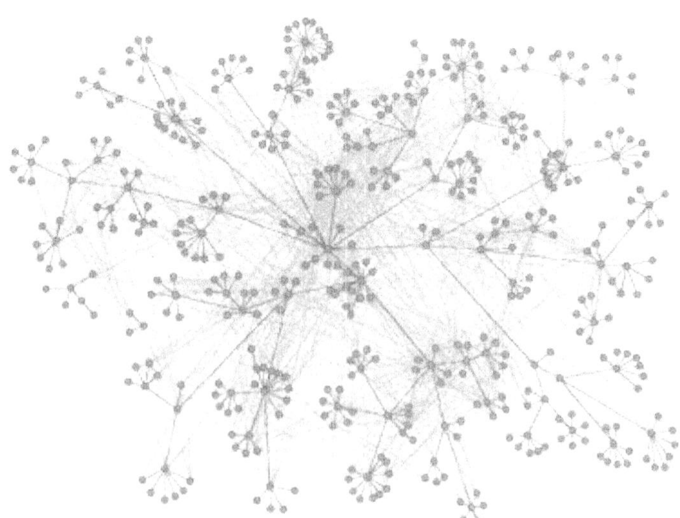

Figure 3.2 E-mail exchanges between company employees [2].

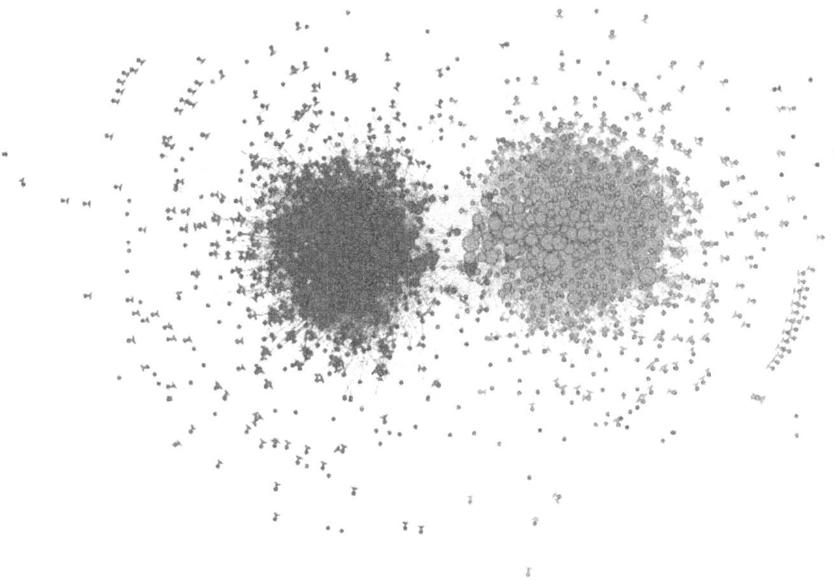

Figure 3.3 Links between web blogs as a form of a large network [2].

they are larger, dynamic, multimode, multiplex networks that may contain varying levels of uncertainty. DNA networks are also more complex than traditional social networks. When comparing DNA and SNA, the most significant difference is that DNA considers the interactions of social features that influence the structure and behavior of networks.

DNA is associated with temporal analysis, but temporal analysis is not always associated with DNA, because changes in networks can be caused by external factors that are unrelated to the social features found in networks, and thus are not always associated with DNA [11].

3.3 Graph

A graph is a common data structure that consists of a finite number of nodes (or vertices) and a connected set of edges [3].

3.3.1 Node, Edges, and Neighbors

A graph is a tool that is used to depict the connections between a collection of objects. A graph is made up of a collection of items, known as nodes, which are linked by links, known as edges, to form a network of connections.

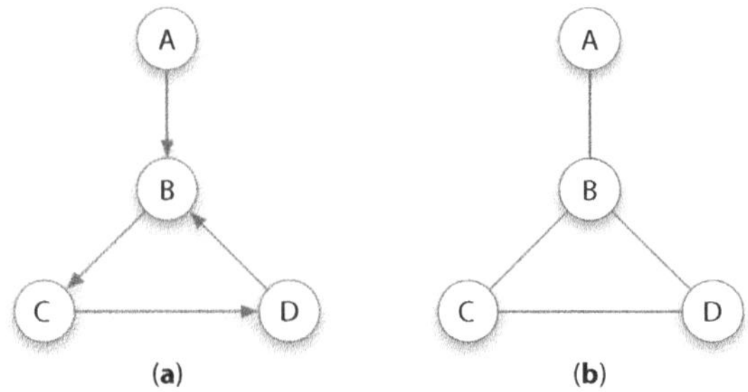

Figure 3.4 (a) Directed graph with 4 nodes [4]. (b) A graph with 4 nodes [4].

A good illustration of this is seen in Figure 3.4(b), which consists of four nodes denoted by the letters A, B, C, and D, which are each linked to the other three by edges. In addition, nodes C and D are linked by an edge as well. It is required that two nodes be linked by an edge to refer to them as neighbors.

Drawing a graph is shown in Figure 3.4 by using a series of dots to represent the nodes and a line to connect each pair of nodes that are connected together by an edge.

A directed graph is made up of a collection of nodes (a node-set) and a collection of directed edges (a directed graph). The directed edges are connections between nodes in which the direction of the connection is essential.

Figure 3.4 depicts a graph in which the nodes are represented by boxes, and the edges are represented by arrows (a). An undirected graph may be used to bring attention to the fact that a graph is not directed when we wish to make a point about it [4].

3.3.2 Small-World Phenomenon

It is a fundamental statement regarding the abundance of short routes in a graph whose nodes are people, with connections linking pairs of individuals who are familiar with one another, which is a significant subject in social networks. It is also a topic on which the feedback between social, mathematical, and computational problems has been very fluid, as a side note.

To trace short paths through the United States' social network, Milgram conducted a series of experiments in the 1960s in which he gave participants the option of forwarding an unsolicited letter to a "target person" in the Boston area. The only restriction was that each participant could advance the letter only by forwarding it to a single acquaintance. Milgram found that the average length of a completed chain was six links long on average, according to Milgram's research.

We are baffled as to why a social networking site would have so short paths. More recently, applied mathematicians Duncan Watts and Steve Strogatz proposed thinking about networks with this small-world property as a superposition: a highly clustered subnetwork consisting of the "local acquaintances" of nodes combined with a collection of random long-range shortcuts that aid in the production of short paths to help with the production of short paths.

Watts and Strogatz investigated the following fundamental model system as a supplement to empirical studies of social, technological, and biological networks:

Construct an ad-dimensional lattice network and link it to a restricted number of long-range connections that originate at each node and end at destinations that are selected evenly at random to get started.

Similar to how many real-world networks are characterized by local clustering and short routes, a network created using this superposition will be characterized by local clustering and short pathways as well [5].

The structure of the small-world phenomenon is an excellent illustration of what occurs during social media, which results in the formation of massive networks that must be managed by massive components and scientific software tools. Providing models that are used as a small type, or more precisely, a sample of that massive network in a smaller size is one method of analyzing and obtaining approximate data from the network.

Whereas this reasoning is mathematically sound, how much it reveals about actual social networks is unknown. There are mathematical methods for approximating the relationship and possibilities between nodes. The clustering coefficient indicates the possibility of a relationship between two nodes in a model. The clustering coefficient values fall within the range [0,1]. If the clustering coefficient indicates that the probability of a relation existing between nodes is 1% or 100%, this indicates that connections exist between all nodes in the model; otherwise, the other values in the interval are used to explain the rate of probabilities of connection exists between nodes. For example, a rectangle with three vertices has the clustering coefficient of one that means all the three vertices while being neighbors of each other they have a direct connection to one another [6].

3.4 Scale-Free Network

Scale-Free Network or real-world network is an alternative to traditional network models, which provide the setting to simulate large networks. One of the main differences between a scale-free network and the small-world network is the two models have different fundamental properties, where the latter model has two critical properties:

1. The network grows over time (Figure 3.5).
2. Vertices and edges are willing to join other vertices and edges (preferential attachment) (Figure 3.6).

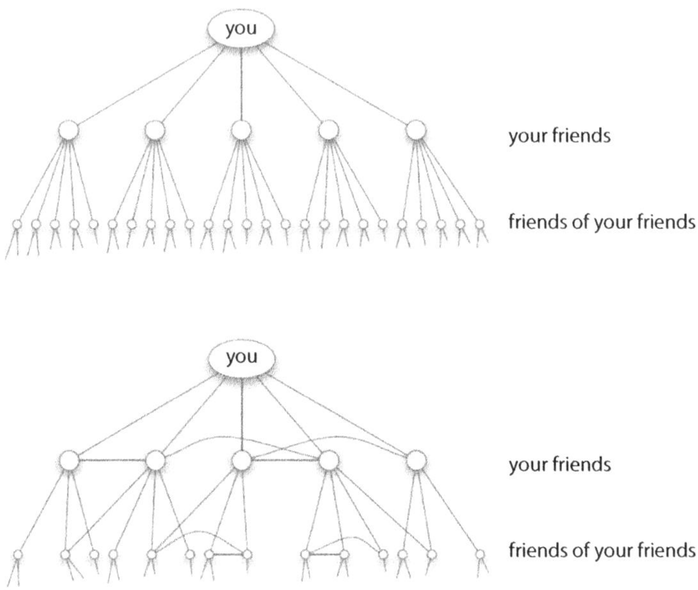

Figure 3.5 Social networks quickly expand to reach many people [6].

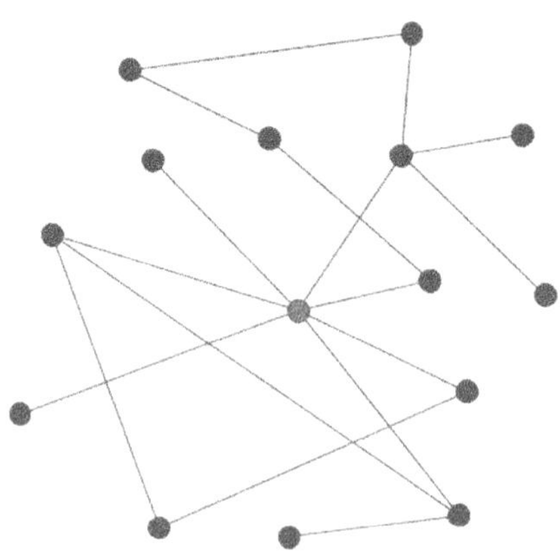

Figure 3.6 A small network that has one highly connected node, or hub [7].

Scale-Free Network (Figure 3.7) contains a large number of nodes with a high degree of the neighborhood, which is the so-called hub node, as well as some nodes with a low degree of the neighborhood, where the higher number is with the nodes with the lower degree or neighborhood rate. Small-world network modeling would be unable to replicate the particular behavior when it is situated under certain circumstances, and SFN simulation will be used as a result.

The researchers studied the network to see which of the aforementioned factors are significant in describing the characteristics of the small-world network. The number of theories, Watts-Strogatz model, Newman-Watts model, Highly Connected Extra Vertex Model (HCV), and Dorogovtsev-Mendes-Samukhin model, Barabasi-Albert model, Krapivsky-Rodgers-Redner model, Vazquez model, Davidson Ebel Bornholdt model, concerning the small-world phenomenon consists of numerous models. However, in SFN, the only model is the Barabasi-Albert model, Dorogovtsev-Mendes-Samukhin

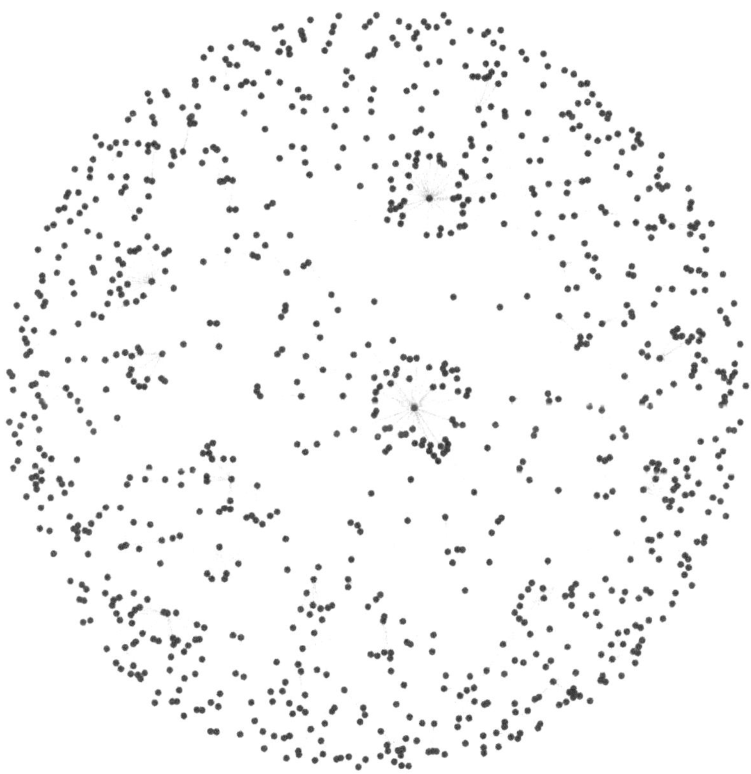

Figure 3.7 Scale-free network in vast size where the hub nodes could be distinguished from the other nodes [8].

model, Krapivsky-Rodgers-Redner model, Vazquez model, Davidson Ebel Bornholdt model. These models have different functionalities, yet they are all employed for different network behaviors.

In other words, in the real world, email systems could serve as an example of this type of network model. To begin, each email address is a node in the system, and when an email is sent from the sender and received by the second person, an edge is produced. Because of the existence of the graph of emails, the scale-free network will be formed, which, because of the peculiarities of this model, can pass both the graphical representation of emails and the algorithm that governs how many emails are in each user's inbox.

As far as large-scale-free networks are concerned, there is no risk of random attacks on the nodes. Because they are reliant on the few highly linked nodes, they are especially vulnerable to targeted assault. However, the network's connection is unlikely to be impacted even if a single hub fails since more hubs exist. Think of it this way: Being linked means that each node is connected to every other node. You would guess that the network structure is an essential problem when it comes to the study of cybersecurity [9].

3.5 Network Data Sets

Large-scale network research has accelerated in recent years, owing in large part to the increased availability of large, detailed network data sets. Additionally, it is worth noting that, while determining why you are interested in studying particular network data sets is a good place to start, there are several other, distinct reasons to do so.

Second, you may be concerned with the source domain, in which case you may be interested in both the big picture and the fine details of the data. Another possibility is that you are using the data set as a proxy for a related network, which is difficult to quantify.

Likewise, you could be experimenting to identify properties that are frequently shared across multiple domains, and thus finding a similar result in unrelated scenarios could support the claim that these properties are universal, with possible explanations that are not domain-specific.

Although a researcher may investigate a variety of reasons, including curiosity, practicality, and intellectual curiosity, they are frequently all active concurrently, to varying degrees. However, at a more detailed level, the researchers wanted to learn more about the dynamics of instant messaging. Finally, it is critical to consider the data sources available on large networks.

For instance, suppose one wishes to examine the social network of twenty individuals (e.g., the people in a small company, fraternity, sorority, or karate club—Figure 3.1). The method used in this case is to interview each of those people and learn more about their social networks. To investigate the interactions of 20,000 individuals or 20,000 unique nodes of any type, we must be more opportunistic in our data gathering methods. In the absence of the ability to go out and gather everything personally, we must examine situations in which the data have already been collected for us in some significant manner. Given this knowledge, take a minute to investigate some of the most significant sources of large-scale network data that have been utilized for research purposes in the past. The resultant list is not comprehensive, and the categories are not clearly distinguishable from one another. As a result, it is very difficult to identify a single data set, and many of its characteristics may be derived from a variety of different data sets.

- **Collaboration graph,** It keeps track of who works with whom in a particular environment; in this case, actors and actresses appear in a film, as a result, a collaboration edge will be formed between them, like a co-authorship between two authors who are working on the same article or a book to publish. This means that every author or actor and actress will be formed as a node on the graph and that cooperation in a similar role will result in them forming an edge over each other.
- **Who-talks-to-Whom Graph,** Researchers have been studying a particular kind of graph known as a call graph, in which each node represents a phone number and where there is an edge connecting two nodes if they have interacted by phone call for a certain length of time. A mobile phone with short-range wireless technology, when used near other mobile phones that use the same technology, can also find devices that have a range of only a few feet. To construct graphs showing the physical proximity of subjects, researchers must give subjects devices and record their traces. The ability to construct "face-to-face" graphs, which show the relative physical positions of each participant, and where a node in the graph represents a person who is carrying one of the mobile devices, allows them to conduct their research more effectively. In the graph, an edge is formed by two linked nodes for every observed observation period that has been observed. Always keep in mind that nearly all of these kinds of data sets include nodes that represent customers, employees, or pupils of the business that manages the data. An individual who falls into this category generally has high expectations of privacy, even

if they might not appreciate how easily details of their behavior can be reconstructed from digital traces such as the digital records they leave when communicating by email, instant messaging, or phone. Because of the fact that research is performed in a certain way to safeguard the privacy of people who are included in the data, this kind of study will usually be restricted to this particular type of data alone. Also, a range of other issues concerning the protection of personal privacy has become highly relevant, both for organizations that are using this data for marketing and also for governments that are looking to use it for intelligence gathering.

- **Networks in the Natural World,** A wide variety of graph topologies is also present basically in natural science, and there has been a significant amount of effort put into the study of numerous distinct kinds of biological networks. At three distinct sizes, we have chosen three instances: beginning at the societal level and progressing all the way down to the molecular level. An early example of this concept is the food web, which depicts the interaction between individual species and their consumers. The food web shows an association between a given species and its consumer, which can be a single-species network or a two-way link between two different species. Reasons to understand food web structure in the form of a graph include being able to reason about situations like cascading extinctions, when specific species become extinct, which causes their food source species to become extinct, and which can lead to additional species extinction, especially since those who relied on the first-mentioned species for food were the next ones to go. There is a large amount of research done on another biological network that has been extensively explored: The term "neural connectivity" refers to the arrangement of neural connections inside an organism's cerebral cortex. Neurons are represented as nodes, and each edge represents the connectivity between various neurons. For the *C. elegans* model, which consists of a relatively simple 302-node brain network with around 7,000 edges, we have a very full view of the global brain architecture. However, acquiring a detailed picture of a more complex brain network is just beyond the current state of the art. Despite this, a considerable new understanding has been attained through a study of the organization of various modules within a complex brain, and how they connect. The last example of a network system that constitutes a cell's metabolism is the set of networks making up that system. These networks are extremely complex, and many other approaches to defining them exist. Generally, the nodes represent components that each perform a specific role in a metabolic process, and the edges represent chemical interactions between those components. There is a great deal of hope that studying these networks will reveal the intricate reaction pathways and regulatory feedback loops that take place inside a cell and that this will lead to the development of "network-centric" approaches to treating pathogens that alter their metabolism in specific ways [10–12].

3.6 Conclusion

As a result, the primary challenges that these massive data sets will face are storage capacity, data processing, and data analysis models, necessitating the development of solutions that could be theoretical or software-based, as well as physical or hardware-based. On the one hand, the theoretical or software aspect refers to mathematical equations and the applied software systems that each network model uses to theoretically analyze and program the network; on the other hand, the physical aspect refers to digitalized equipment, such as server devices, hardware storages, and new modern technologies that handle the physical part.

All of these structures have a feature and characteristic, they have to function and analyze the specific network structure. Knowing that understanding what is vital is about properly applying the correct model to the specific network.

As the outcome, handling real-world network data sets would be provided by the proper understanding of the graph theories, as well as understanding network behavior and applying the right model to analyze and model the behavior.

References

1. Marion Neumann, Analysis of network data, The Fekete website, *Handling Large Datasets*, Washington University in St. Louis, Missouri, United States, Fall 2019, http://www.fekete.com/san/webhelp/welltest/webhelp/content/html_files/procedures/preparing_data_for_analysis/handling_large_datasets.htm.

2. Easley, D. and Kleinberg, J., Networks, Crowds, and Markets, Reasoning about a Highly Connected World, Chapter 1, pp 1–5, [online], https://www.cs.cornell.edu/home/kleinber/networks-book/networks-book-ch01.pdf, published by Cambridge University Press 2010.
3. What is a graph (Data structure) by Edpresso team. https://www.educative.io/edpresso/what-is-a-graph-data-structure.
4. Easley, D. and Kleinberg, J., *From the book Networks, Crowds, and Markets: Reasoning about a Highly Connected World*, pp. 23–24, Cambridge University Press, Cambridge, England, United Kingdom, 2010, http://www.cs.cornell.edu/home/kleinber/networks-book/.
5. Kleinburg, J., The small-world phenomenon and decentralized search. *SIAM News*, 37, 3, April 2004.
6. Easley, D. and Kleinberg, J., *From the book Networks, Crowds, and Markets: Reasoning about a Highly Connected World*, pp. 612–614, Cambridge University Press, Cambridge, England, United Kingdom, 2010.
7. Small network with a hub. *From Math Insight*, http://mathinsight.org/image/small_network_hub.
8. An Image from the Flicker website, https://www.flickr.com/photos/sjcockell/8425835703
9. From the future learn website, Scale-Free Networks, https://www.futurelearn.com/info/courses/complexity-and-uncertainty/0/steps/1855. © University of Southampton 2017. https://www.futurelearn.com/info/courses/social-media/0/steps/16046
10. Easley, D. and Kleinberg, J., *From the book Networks, Crowds, and Markets: Reasoning about a Highly Connected World*, pp. 40–44, Cambridge University Press, Cambridge, England, United Kingdom, 2010, Complete preprint on-line at http://www.cs.cornell.edu/home/kleinber/networks-book/.
11. Dynamic network analysis, https://en.wikipedia.org/wiki/Dynamic_network_analysisFind; https://www.google.com/search?as_eq=wikipedia&q=%22Dynamic+network+analysis%22; https://www.google.com/search?tbm=nws&q=%22Dynamic+network+analysis%22+-wikipedia&tbs=ar:1; https://www.google.com/search?&q=%22Dynamic+network+analysis%22&tbs=bkt:s&tbm=bks; https://www.google.com/search?tbs=bks:1&q=%22Dynamic+network+analysis%22+-wikipedia; https://scholar.google.com/scholar?q=%22Dynamic+network+analysis%22; https://www.jstor.org/action/doBasicSearch?Query=%22Dynamic+network+analysis%22&ac-c=on&wc=on JSTOR (April 2009).
12. Galety, M.G., Saravana-Balaji, B., Saleem-Basha, M.S., OSSR-P: Ontological service searching and ranking system for PaaS services. *Int. J. Adv. Trends Comput. Sci. Eng.*, 8, 2, 271–276, 2019.

4

Cascading Behavior in Networks

Vasanthakumar G. U.

Department of Computer Science Engineering, Nitte Meenakshi Institute of Technology, Bengaluru, India

Abstract
This chapter elaborates on determining the Cascading Behavioral Pattern of Social Network users. The data available in social media are usually the user-generated content, comprising of images, text, video, and so on, and are unstructured. Social networks users are of various types who use the platform for varied reasons. Here, the influencers are a type of social network users, who influence other users on various backgrounds. The contact generated by users like videos, posts, images, and so on, are the major components used for influencing. The format or pattern of influence may depend on various factors. For profiling the user in social networks, the parameters like user actions, patterns of activities, behavior, posts make a major contribution because these variables characterize the users. Businessmen take various steps to promote their products using the behavioral pattern of users in social networks. The combination of machine learning algorithms and natural language processing together works as a backbone to understand the text content of data and the user behavioral pattern in social media.

***Keywords*:** Cascading behavior, online social networks, pattern of activities, profiling, unstructured data

4.1 Introduction

Online social networks (OSN) have brought together numerous individuals across various locations of the world generating a tremendous amount of data every day. The data are getting generated at a faster pace online, and social media is one of the main and highest data-producing

Email: vasanth.gu@nmit.ac.in

Mohammad Gouse Galety, Chiai Al Atroshi, Bunil Kumar Balabantaray and Sachi Nandan Mohanty (eds.) Social Network Analysis: Theory and Applications, (51–62) © 2022 Scrivener Publishing LLC

sources at present time. These sites provide a platform for its users to express their opinion, share knowledge, create impact, promote business, and so on. The OSN facilitates the creation and diffuses the information in a cascaded way. With the ongoing social, political, cultural, tourism, and other activities in the real world, the topics of discussion vary in social networking sites by the users. The online data available in social media are usually user-generated content, which contain images, text, videos, and so on, and are unstructured. In recent days, OSN has gained huge attention from the business point of view where industries consider knowing their audience as the key to their success. Many companies provide various tools for customers to engage and discuss their experience on products, creating interactive community, and get socialized with visitors as well as with their customers.

4.1.1 Types of Data Generated in OSNs

Usually, the data generated in OSN are unstructured and are rarely structured. For the analysis or aggregation of such data to use for various purposes, it mandates to structure and shape the data content. Like earlier times, simply categorizing networks according to their functionalities: for example, short text content for Twitter, videos for YouTube, and so on, no longer holds good for any research or marketing strategies. The different networks provide various rich features in respective domains to enrich users' experience and to widen the domain capabilities. It is high time to use new techniques to break down the social network data and to categorize them based on various terminologies, like content analysis, content flow, location, links, followers/following, interests, history, and so on. Texts, videos, audios, discussions, forums are few examples for categorizing the unstructured data post structuring.

4.1.2 Unstructured Data

Any data that are not arranged as per the requirements of some schema is unstructured. Multimedia, as well as text data, may be considered to be a common type of unstructured content. You may also find many business documents, e-mail messages, videos, webpages, photos, as well as some audio files to be unstructured, which contain a lot of information for making business decisions. Unstructured types of data can have internal structural elements. Simple content searches can be performed on textual unstructured data. Traditional analytics tools are optimized for highly structured relational data, so they are of little use for unstructured sources, such as rich media, customer interactions, and social media data.

Big data and unstructured data often go together and recent tools have been developed to analyze the unstructured sources. Social media scraper is an automated web scraping tool that is used to extract social media data. Various social media analytics are available in the market amongst which few are free and others are paid. Application Programming Interfaces (APIs) were the traditional ways of extracting social data, but now the volume of data through APIs has reduced in large. Apart from web scraping, there are other methods like manual approach, web services, and automated approach so on available to fetch the data from social media.

4.1.3 Tools for Structuring the Data

To understand the beneath meaning of data being generated in the OSN, it becomes an integral part to structure the data using various algorithms/procedures and to shape them. Unstructured data cannot be fit into any framework, it is just a quantitative idea and hence structuring the data is a mandate, which results in qualitative and valued data. The similar contents are grouped to form clusters which reveal the clue of what content that data would be.

RapidMiner, KNIME, Google and Excel sheets, Power BI, Tableau are some of the example tools used for structuring the data of OSN. These pretrained tools can extract keywords, group similar words, performs sentiment analysis etc., to break the unstructured data into a structured format and to shape it. Some of the tools are very interactive where it customizes the user needs by even connecting to third-party applications and provides results in form of graphs, excel, so on. These structured data also helps in drawing the pattern and content flow of data in social networks.

4.2 User Behavior

Profiling users are directly dependent on their behavior, actions, involvement, and pattern of activities in various social networks. Businessmen take various steps to promote their products using social networks, amongst which unethical users are hired who can influence others. Through social networks, they even try to improve the market value of the products unethically. Social network users are of various types who use the platform for varied reasons. Here influencers are the type of social network users, who influence other users of various backgrounds. The contact generated by users like videos, posts, images etc., are the major components used for influencing. The format or pattern of influence may depend on various factors. The Geo-tagging model can be used to identify the locations of photos posted on social networks by users based on the analysis made on the content of user photos.

4.2.1 Profiling

Significant data that might reflect the user interest over the various subject matter is termed as User Profile, whereas user profiling is a process of characterizing them from their behavioral patterns. As and how social network usage increases, identification of its users and profiling them become prerequisites for various security reasons. Profiling OSN users based on their various activities gives us an insight into their behavioral aspects for utilizing their potentiality positively. Profiling OSN users requires various activities and actions performed either on their own or by the influence of others in the network. Though the data aggregation of cross-linked sites is challenging, it becomes a mandate data to profile users of multi-social networks.

4.2.2 Pattern of User Behavior

It is observed that various activities carried out by the information creators pose threat and profiling such users by analyzing their behavioral patterns is needed. Such users, also called intruders influence other users for carrying out various unethical activities. Multi-social networks are also to be analyzed to determine the users' behavior in multiple social networks. For the business to plan strategies and to improve the business through OSNs, make use of users who can propagate the information in a cascaded manner for good reasons like trading and marketing.

In OSN, to promote various businesses and increase their product market value, take the help of predictors, who identify the active users by performing various actions for advertisements. A method has been developed [1] for ascertaining the value of predictors as incorrectly predicting them may lead to a loss in their business. Therefore, tracking the user activities will allow us in characterizing them, which in turn contributes to profiling that user. ClickStream [2], a statistical model based on the URL clicks of the user is developed through which the pattern that the user has followed can be identified.

Social interactions and their patterns are identified using a social network aggregator. A user might have accounts in multiple social networks and may perform various activities. To analyze and track the pattern of user behavior in various social networks appropriately algorithms are used, which manually classifies the user statistics and their activities. By analyzing the image posts of users, Geo-tagging is achieved. GPS mechanism is used by a Geo-tagging system [3] which is developed to locate the location of images uploaded in social networks. User interest is subject to change as time elapses. This is proven by developing a Mobile Social Networks algorithm [4] based on Activity Prediction.

One-to-one influence in OSN is common and apart from that, the other trend is to influence groups of users termed as Community Influence. Here, one individual influences the entire community and one community (one user in that community) influences other communities through posted photos, videos, messages etc. The online social users and their respective communities are identified [5] and their unusual activities are detected by a system. The system [6] also depicts the behavior of users in social networks. Innovation is made with the integration of a GIS system [7] to perform statistical analysis and dynamic data visualization. Users behavioral pattern is analyzed [8] by developing a prototype. This prototype uses the details of users' interest in accessing the available information on the internet.

The network flows and their pattern is analyzed to identify the malicious behavior of users by feeding the data to a machine learning-based classification process [9]. There are like and unlike-minded people in social networks, Proximity of nodes [10] in the network is measured for analyzing and identifying social like-minded users. To identify the behavior of students while learning the short courses online, Educational Data Mining (EDM) [11] is used. The behavior patterns are extracted and students are grouped based on their patterns of behavior.

4.2.3 Geo-Tagging

Geo-tagging is a method adopted in adding the location details like latitude and longitude values to multimedia content like websites, audios, videos, and photographs. This can even be applied to tweets or status updates on social media. This Geolocation data from various sources provide insight into social influence and individual behavior. Most social networks and their related applications utilize geo-tagging to track the location of their users and in turn their behavioral patterns. Most common is geo-tagging of the pictures caught by their camera and associating it with the location, which will occur after it is posted online. The correlation between the social relationships and individual-specific patterns of users can be analyzed using the data being captured. The similar activity choice and geo lifestyle patterns between two individuals who are socially connected can be extracted by processing their social network data.

4.3 Cascaded Behavior

Cascading Behavior in Networks can be described as "The influence of one or more users on others for decision making when all are connected in the same network". Here actions of one user may create a trend and influence others who are connected to perform the same or similar actions over the network. To identify the pattern of user activities, actions and their behavior in social networks, there are multiple statistical models available. For profiling users in social networks, the parameters like user's actions, patterns of activities, behavior, posts make a major contribution since these variables characterize the users. Though the variables for profiling users are captured accurately, profiling users cannot be concluded just by extracting these details from a single social network. Instead, the same extraction needs to be performed in various social networks related to the same user(s) using various methods/models or techniques. The combination of machine learning algorithms and natural language processing together works as a backbone to understand the text content of data and the user behavioral pattern in social media.

4.3.1 Cross Network Behavior

Although online social networking sites have brought diverse people around the world to be socially connected with one's convenience, it still provides a threat to privacy protection and information leakage. Privacy protection and disclosure of information are two important factors though seem to be a single element in ongoing research topic on a security basis. Literature shows that many users fail to perceive privacy risks and fail to behave following potentials risks provided in form of awareness. A user, in general, may have multiple social network accounts in either the same or different networks. We ignore users with multiple accounts in the same network for now and concentrate on users having accounts in multiple social networks. Commonly, users try to have the same Username and profile pictures to be the same over multiple networks to maintain visibility to the other individuals or groups. Some malicious users showcase completely different personal information over networks. As security is a concern, few users avoid sharing some personal

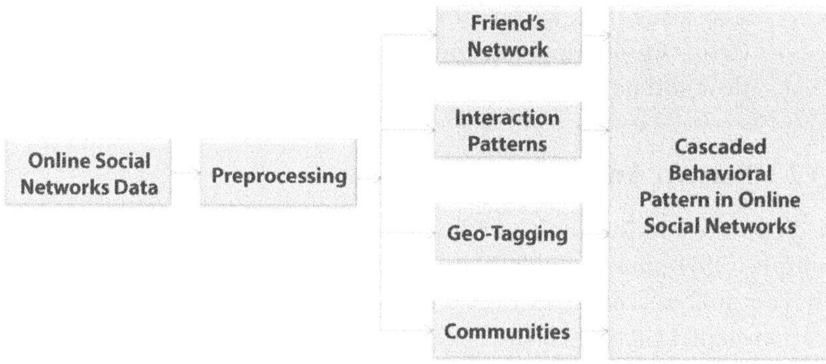

Figure 4.1 Aspects of cascading behavioral pattern.

information on social networks. Figure 4.1. shows various aspects of Cascading Behavioral Pattern in OSN.

Study on behavior reveals very important key points, which includes the privacy breach and malicious accounts in OSNs. These malicious accounts generate unwanted messages and an unhealthy environment within legitimate users. Henceforth, identifying and blocking such malicious users is very important in ensuring a good user experience [12]. As discussed, users often hold accounts on multiple sites, cross-site information aggregation is challenging. Cross-linking the multiple social network accounts of users helps in analyzing the social activities, concerns and their respective interests, clustered groups, friends of users to profile them. To understand user activities, motivations and characteristics on cross-site linking, asking questionnaires like e.g., asking user why and why not chooses to enable cross-site-linking function, why multiple accounts in various networks without connectivity etc., to users by conducting surveys. When users have not cross-linked their various social site accounts, then the behavior of users can be profiled by considering their respective multiple accounts along with their social content posted, location, pictures, friends, profile first name, last name so on. Cross-linking user accounts in multiple OSNs can be useful in many ways. Some of them are:

- Cross-site linking makes cross-site content posting easy and fast,
- Cross-site linking makes it very easy to build the friend network or groups/social connections,
- Cross-site linking provides more information of a user, beyond that stored on a single OSN site,

- Cross-site linking can provide users behavior pattern,
- Cross-site linking can provide the cascade of information flow and its shape and size etc.

4.3.2 Pattern Analysis

As cross-site linking makes it possible to aggregate information from multiple OSN sites though it is challenging, pattern recognition and behavior analysis can be extracted using these cross-linked sites from the location-centric activities and social interactions of users [13]. A known behavior and description of observable actions of the user in multiple OSNs, which are cross-linked forms a pattern. The related pattern of activities also are observed in cross-linked social networks with the same or similar shape and these are very helpful in extracting the behavioral pattern. Many times, the pattern flow holds as a reference to capture the other formed patterns and to evaluate the users by their actions and activities in respective patterns in cross-liked social networks. Verification of user's personal information shared publicly in cross-linked social networks reveals the fact whether the same information is provided across networks or some discrepancies noted. Some assumptions may hold good in cross-linked social networks that the profile names, profile pictures, basic information, location and interests may be the same. In some cases, the friends and subgroups could also be the same. Content analysis plays a very vital role in extracting pattern recognition and information cascading details in social networks. The pattern of information cascade reveals the interconnection of subgroups, individuals and larger groups. The pattern cascade shape along with its size are the major components in analyzing the behavior in social networks. The cross-site information consistency can be accurately evaluated using the cross-site linking function [14]. Analysis on Foursquare and Facebook social data of users [15] revealed a consistent percentage of the first name field and the last name field as 89.84% and 87.02%, respectively, while that of the gender field is 99.30%. With good consistency of cross-linked social network data of users and frequent cascade subgroups, it is easy to predicts and draws the pattern of information diffusion and pattern of information being generated in OSNs.

Let us consider an example as shown in Figure 4.2. The figure shows a small group in a social network with six nodes A, B, C, D, E, and F. Here, nodes represent the users of real-world in OSN. Assume that all six nodes are following the trend P at first. Later, when a new trend Q comes, it gets adopted by node A. Further, other nodes viz B, D, C, E, and F also start

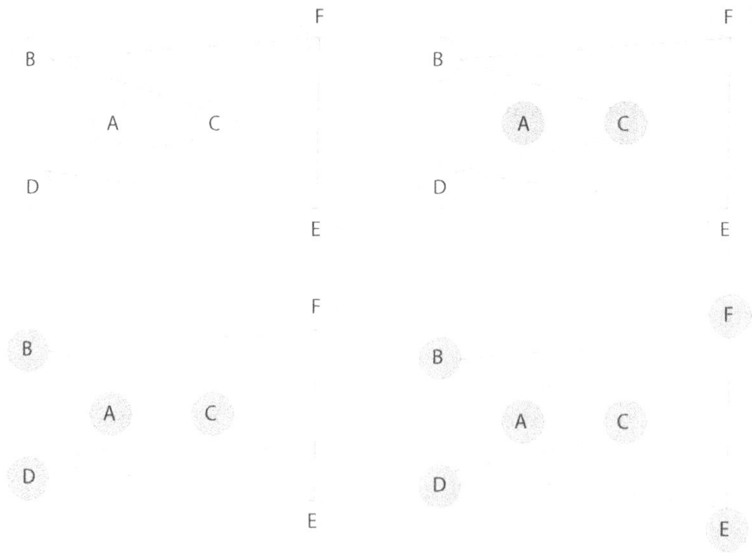

Figure 4.2 Cascading behavior in OSN.

adopting and following the pattern as shown in Figure 4.2., which shows how the new trend Q is adopted by the neighboring nodes depicting the cascade of trend in OSN.

4.3.3 Models for Cascading Pattern

Based on one's research, various models exist, which are self-explanatory in providing the mechanisms followed to find the cascading pattern in OSNs. Many times, these are applicable for real-time scenarios as well. Assume a huge network with millions of users, the information diffusion in such networks would be very quick and spreads enormously across the network and the pattern followed in cascading the information need not be the same. There could be different patterns, flows and trends in the format of cascading, based on the clustering of users and formation of groups and sub-groups in the network.

Legacy Models: The process of adoption of information or trend in the Diffusion model can be classified into two categories namely: *The threshold model* [16] drawn from some probability distribution. With connection weights on the edges of the network, if the sum of connection weights of neighbors of a node that already adopted the behavior is greater than the threshold, then the current node adopts the behavior. In an *Independent*

Cascade Model [17], every time a neighbor node adopts a behavior, the chances of the next node adopting the behavior are huge with a higher probability.

Explanatory Models: A node in a social network can represent a "Real" user of a society. Interactions between such nodes i.e., two users can be represented as edges meaning, the existence of the relationship between two nodes [18, 19]. By considering the huge social network and a piece of information, the dissemination of information along with the pattern, shape, size, subgroups may be captured with Basic Epidemics Models like the SI Model, the SIS Model, as well as the SIRS Model in Social Networks.

Influence Models: The other kind of model in OSN is the Influence model. This model describes the formation of influence by one node to others. Here, influence is of different kinds like one-to-one, one-to-many, one-to-group, e.g., community spread, group-to-group, and so on.

Predictive Models: The spread of information in social networks published by an individual is very quick and vast. The evaluation of such spread can include the parameters like speed, shape, time, pattern, number of nodes etc. As the name indicates, Predictive models are useful to predict the future information diffusion patterns in OSN based on various factors. The future spread of information may be predicted using these prediction models and are useful in applications like product sales, government bodies to control the situation, demand in the society, and so on.

References

1. Goel, S. and Daniel, G.G., Predicting Individual Behavior with Social Networks. *Market. Sci.*, 33, 1, 82–93, 2014.
2. Benevenuto, F., Rodrigeus., T., Cha, M., Almeida, V., Characterizing User Behavior in Online Social Networks (OSNs). *IMC*, 2009.
3. Papagelis, M., Murdock, V., Zwol, R.V., Individual Behavior and Social Influence in Online Social Systems. *HT*, 2011.
4. Gong, J., Tang, J., Fong, A.C.M., ACTPred: Activity Prediction in Mobile Social Networks. *Tsinghua Sci. Technol.*, 19, 3, 265–274, 2014.
5. Angeletou, S., Rowe, M., Alani, H., Modelling and analysis of User behaviour in online community. *ISWC*, pp. 35–50, 2011.
6. Youssef, A. and Emam, A., Network Intrusion Detection Using Data Mining and Network Behavior Analysis. *Int. J. Comput. Sci. Inf. Technol. (IJCSIT)*, 3, 6, 87, 2011.
7. Bianconi, F., Brunori, V., Valigi, P., La Rosa, F., Stracci, F., Information Technology as Tools for Cancer Registry and Regional Cancer Network

Integration. *IEEE Trans. Syst. Man Cybern. A. Syst. Hum.*, 42, 6, 1410–1424, 2012.
8. Paek, H.-J. and Hove, T., *Determinants of Vertical and Horizontal Online Health Information Behavior*, IEEE, 2014.
9. Gokcen, Y., Foroushani, V.A., NurZincir-Heywood, A., *Can we identify NAT behavior by analysing Traffic Flows?*, IEEE, 2014.
10. Liben-Nowell, D. and Kleinberg, J., The Link-Prediction Problem for Social Networks. *J. Am. Soc Inf. Sci. Technol.*, 58, 7, 1019–1031, 2007.
11. Ratnapala, I.P., Ragel, R.G., Deegalla., S., *Students Behavioural Analysis in an Online Learning Environment Using Data Mining*, IEEE, 2014.
12. Jin, L., Chen, Y., Wang, T., Hui, P., Vasilakos, A.V., Understanding User Behavior in Online Social Networks: A Survey. *IEEE Commun. Mag.*, 51, 9, 144–150, September 2013.
13. Klamma, R., Spaniol, M., Cao, Y., Jarke, M., Pattern-Based Cross Media Social Network Analysis for Technology Enhanced Learning in Europe, in: *Lecture Notes in Computer Science*, 2006.
14. Chen, T., Kaafar, M.A. *et al.*, Is More Always Merrier? A Deep Dive Into Online Social Footprints, in: *Proc. of ACM WOSN*, 2012.
15. Chen, Y., Cao, Q., Hui, P., *Understanding Cross-Site Linking in Online Social Networks*, 2014.
16. Granovetter, M., Threshold models of collective behavior. *Am. J. Sociol.*, 83, 6, 1420–1443, 1978.
17. Goldenberg, J., Libai, B., Muller, E., Talk of the network: A complex systems look at the underlying process of word-of-mouth. *Market. Lett.*, 3, 12, 211–223, 2001.
18. Li, M., Wang, X., Gao, K., Zhang, S., A Survey on Information Diffusion in Online Social Networks: Models and Methods. *Information*, 8, 4, 118, 2017.
19. Kumar, A., Galety, M.G., Nuru, M., Adam, T., WCBAODV: An efficacious approach to detect wormhole attack in VANET using CBAODV algorithm. *Int. J. Adv. Trends Comput. Sci. Eng.*, 8, 1, 20–25, 2019.

5

Social Network Structure and Data Analysis in Healthcare

Sailee Bhambere[1,2]

[1]*BDS, MPH, Harvard University, Boston, Massachusetts, USA*
[2]*Senior Clinical Pathways, Health Innovators Inc., Boston, Massachusetts, USA*

Abstract
With the development of Internet technologies, social networks attain an indispensable part in the day to day life of human beings because it enables better understanding of the perplexity of the interconnectedness. One methodology that has been predominantly helpful in finding covered up connections, associations, and patterns of complex frameworks through numerical and graphical strategies is social network analysis (SNA). This methodology has become progressively useful for healthcare, specifically as a large number of issues connected to the healthcare frameworks with dynamic entertainers that connect with one another and show emerging complex practices. It acts as an assistive medium for both the patients and the healthcare advisors, as medicine alone is not sufficient for the recovery of patients. The tremendous usage of social media states that the people depend mostly on online available information rather than advertisements for making purchases. The assessment of cost and quality rating of the public leads the way for others to know about better healthcare advices. The health options provided by the social media help the public to be aware about the health-related issues. This chapter provides a deep insight of SNA and its applications in the healthcare system.

Keywords: Social media, SNA, healthcare organization, electronic health records (EHR), SNA application in healthcare

Email: saileebhambere@yahoo.com; saileebhambere@hsph.harvard.edu

Mohammad Gouse Galety, Chiai Al Atroshi, Bunil Kumar Balabantaray and Sachi Nandan Mohanty (eds.) Social Network Analysis: Theory and Applications, (63–82) © 2022 Scrivener Publishing LLC

5.1 Introduction

The escalating occurrence of nontransmissible diseases in conjunction with the increasing population of people in various parts of the world has raised the cost of healthcare provision and rate of mortality in the present era. Nontransmissible diseases, like cancer, cardiovascular diseases (CVDs), respiratory diseases, and diabetes, are noticed to be the major reasons for disability, reduced life quality, and even death, leading to a raise in the cost of healthcare provisions [1]. Despite this huge expenditure, the organizations of healthcare are expected to provide medical services of high quality at reduced costs to their patients. However, payment of high price for medical services alone does not assure services of high quality. Hence, the healthcare organizations are facing a lot of challenges, like increased number of patients, increased price of equipment's engaged in medical practices, and limited financial plan. The organizations of healthcare are expected to utilize advanced technological improvements to provide effective services at reduced costs and increased quality [2].

Hence, the conventional medical practices are now being replaced with proactive approaches that include prognostic, protective, participatory, and personalized (P4) medicine [3]. The electronic health records (EHRs) are attaining more interest with the developments in digitization and technologies of medicine. This helps the healthcare physicians and providers to analyze the potential diseases at its early stage in such a way to monitor the health status of the patients. As most of the diseases are preventable at an earlier stage, monitoring the lifestyle of the patients helps in obtaining the required data about the patients in such a way to attain recovery. Usage of wearable techniques helps in gathering such data in a continuous way for the enhancement of disease monitoring. Healthcare analytics is executed on large amount of data for the timely prediction of the disease to prevent the disease [1].

5.2 Prognostic Analytics—Healthcare

The effectiveness of predictive Medicare analytics can be remarkable when used properly for the diseases, even from common cold to severe diseases, such as cancer, diabetes, sepsis, and so on. The prognostic analytics with the concentration on Medicare is growing rapidly as an efficient mechanism that can authorize proactive, illuminative, and protective treatment options. Using the predictive health analytics on the data can increase the clinical healthcare research. The moment has come to connect and discover integrative character of Medicare and reckoning. Fetching together these domains may enhance the preeminence in decision making and clinical analysis in such a way that it providea additional care to patients by the doctors with the recognition of adverse events at its initial stage, with the visualization of trends and insights in the clinical information. Thus, it is revealed that the healthcare-based decision making shall be rendered by the doctors for which social media plays a major role [4].

5.3 Role of Social Media for Healthcare Applications

Social media, the base of communication from an online medium, allows the interaction and the collaboration of a number of users in a way to share information among them. For instance, the most commonly used social media applications are Twitter, Facebook, Instagram, and so on. These applications involve in sharing pictures and videos to convey their opinions related to business, technology, healthcare, entertainments, and so on. Social media acts an important platform for the effective communication between the healthcare advisors and the patients through the Internet. It acts as an assistive medium for both the patients and the healthcare advisors, as medicine alone is not sufficient for the recovery of patients. The tremendous usage of social media states that people depend mostly on the online available information rather than advertisements for making purchases. The assessment of cost and quality rating of public leads the way for others to know about the better healthcare advices. Figure 5.1 represents the social network in healthcare.

The different responsibilities of social networking in healthcare include the following:

(a) Advertising: Marketing or advertising acts as one of the major skills for the healthcare agencies to be reached worldwide. The social media platform is used to market the mission and the customs of the healthcare organizations in such a way that it will attain the faithfulness of the patients [5].

(b) Information Interchange: Sharing the balanced data publically is the most important function of any social media platform. The specific health-related data of the patients must not be revealed by any corporation without the knowledge of the patient, keeping it confidential yhen. This is the important factor that is needed to be taken into consideration while disclosing the details in social networks. Health ranker website acts as a noticeable example that gives the updates to patients about their health status.

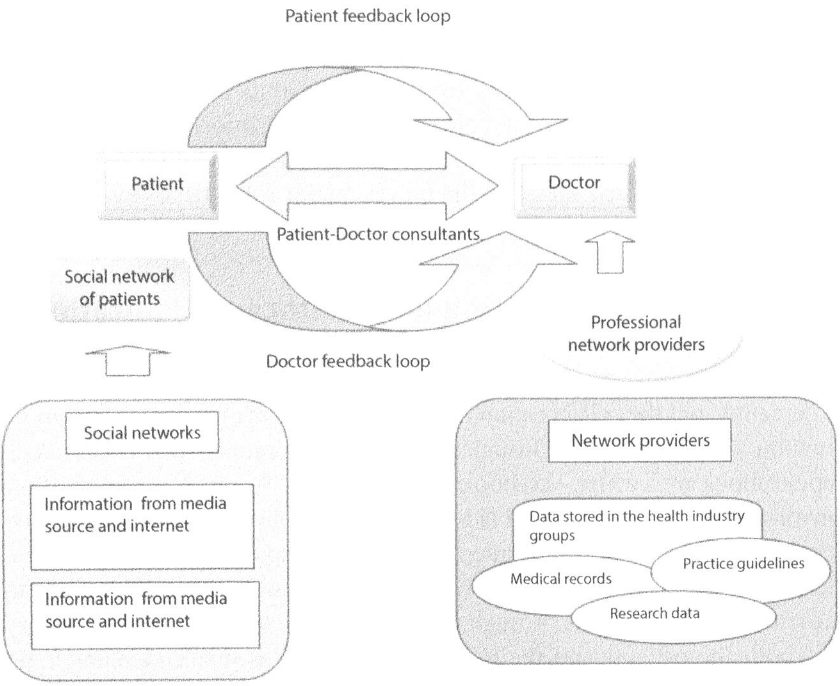

Figure 5.1 Social network showing links and nodes in healthcare.

(c) Research Objectives: The website, like health map, helps the researches of the diseases to analyze a particular disease in detail. The developments in the medical field can be obtained by providing professional connections for the patients, medical students, and doctors through the websites.

(d) Medical assistance: Sites, like DailyStrength, support the patients and the caretakers, where the patients express their experiences and struggles to the caretakers to help themselves with emotional support.

Additionally, the data analysis strategies used in the social media platforms must also be taken into consideration. Whenever the raw data are obtained from the users, the data analysis strategies must involve analyzing the data for the extraction of the significant features from the data to be processed. The aforementioned processes are termed as social media mining. Sentiment Orientation, Opinion Analysis, and Unsupervised Classification are the commonly used social media mining methodologies even though there are various other methods available for social media mining [6].

5.4 Social Media in Advanced Healthcare Support

The healthcare–based applications are quickly developing with an unbelievable speed because of the remarkable increase in healthcare communication through social media. The caretakers and the patients are allowed to communicate with each other whenever necessary with less expense with the introduction of new marketing services, wellness programs, and developments in patient care. The healthcare Hash-tag Project is an important project that leads Twitter more accessible to the users for healthcare communication. The health options provided by the social media help the public to be aware of health-related issues. Companies must come forward to provide health-related data to the public with the ability to pay attention to the requirements of the patients in the social media. The patients, thus, may be capable of diagnosing the diseases by themselves, without the need to visit a doctor [7]. Some of the important statistics about social media in Medicare applications are stated below:

- In a survey, the users of about 90% within the age limit of 18 to 24 years have indicated their trust on medical statics in the social media sites is shared by the public.The healthcare firms of about 31% have detailed guidelines for social media in written form.
- Over 90% of the smart phone users possess at least one healthcare application related to diet and exercises.
- People of about 41% stated that the choice of their doctors or hospitals may change with the introduction of healthcare applications in social media.

The abovementioned points explain the need for social media applications in the healthcare industries. The hospitals promoted with social media are more likely to be developed as inactive hospitals among the public. Doctor Anna (artificial intelligence doctor) and Health 2.0 are some instances of emerging applications in online healthcare.

5.5 Social Media Analytics

The analytics of social media helps in taking advantageous decisions through considering the data from the social media, through some analytical tools. The statistics from the social media is carefully analyzed to predict the information either as positive or negative, and it is an important factor to attain a beneficial social media strategy through the analytical tool. Based on the provenances of social media, the systematic media may vary in such a way to produce accurate predictions. Followerwonk for Twitter, Iconosquare for Instagram, and Quintly for Facebook are some of the effective analytical tools, which are costless, but with certain limitations. For instance, the analytics for about only three Facebook pages can be attained while using the Quintly tool [8]. Lots of websites in social networking encourages the researchers to concentrate on analysis. The Advanced Programming Interface (API) of Twitter allows the developers of the analytical tool to consider the tweets to be analyzed to find the anonymized data. This helps the pharmacists to identify the mistakes with the criticizing tweets made on a particular medicine in such a way that flaws are rectified. Reviews shared online on the social media are seen to be true and analyzing them thoroughly helps the analytics to understand the thinking of people regarding any firm, which helps in improving the quality of the product or service for the firm.

5.5.1 Phases Involved in Social Media Analytics

The three important processes of social media analytics are the trapping, comprehension, and attending [9]. The detailed explanations of the processes are given below:

1) Gathering of data: In the first step, the corporation recognizes the data related to their products, brands, services, and so on, through the social media platforms. Increased amount of information is attained even from a single social media platform. The collected data need to be processed to move to the next step. The processes include trapping and connecting the information from various provenance, preserving the obtained data in a single mart of data in compiled form, developing data models of certain types, extraction of significant parts of data, elimination of unwanted data, and the removal of noise from the data to attain meaningful analysis on the data.

2) Data perception: When the statistics are obtained and packed in an exclusive mart, the analysis of the required information is given in stars. "Understand" is a significant step of data analytics in the social media network, because this step gives the organization an idea or feedback about the feelings of public on their services or products. After performing the data analysis based on certain indices, the prediction of the customers is done toward the initiative to purchase the product or services. There are a number of indices framed to analyze the perspective of the patients. Some of the best indices for analysis are the number of people sharing their opinions about the product or service, volume, and so on. The step "understand" plays a vital role in the next step present.

3) Data visualization: The results from the understand step are analyzed and presented as a summary to the pharmacist or the organization in an understandable format in the present step. A number of visualization methods, like visual dashboard, are used to present the summarized data to the organizations in different types of graphs.

5.5.2 Metrics of Social Media Analytics

The metrics of social media is very important in the evaluation of the activity of social media on a particular organization. The selection of the significant metrics among hundreds of metrics is a crucial challenge. The value of the brand of an organization is necessary to be monitored continuously for the caretakers using the metrics. Some of the commonly used metrics in healthcare organizations are explained below:

a) Volume: Volume acts as a simple metric that helps in the analysis of a number of messages from a particular brand of an organization and the number of people that posted the messages to their known circle.

b) Reputation: The spread or the popularity of an activity in social media is termed as reach or reputation. However, to make this metric to be effective, some other metrics must be combined with it. Generally, the metric reach is used as a denominator term in the equations of measurements related to social media.

c) Dedicated users: Dedicated users are also the best metric that involve in finding the people engaging on activities detailing the conversation about the products or services to recommend it.

d) Dominance: The dominance metric helps in selecting a person with high influence on public to be approached to broadcast a particular product.

e) Metrics Evaluation: To be superior, the performance must be higher for a particular organization compared with the participants. Demographics Pro [3] is an analytic appliance that aids the dealers to obtain the necessary social media specification with social ventures in popular websites, like Facebook, Instagram, and so on. This helps the marketers to focus on the platforms for maximum benefit [10].

5.5.3 Evolution of NIHR

The development of the innovation theory prevail the structure for the fusion of technical advancement in the Medicare system through the social networks [11]. The implementation of the innovation theory is widely preferred in the UK because of the great extent to worries about the absence of take-up, interpretation into training, and the information on the impacts of mediations in medical services. The contribution of the National Institute for Health Research (NIHR) in the medical care services and the organization with the recent technical development due to the intervention of the Collaborations for the Leadership in Applied Health Research and care (CLAHRCs) draw the attention of healthcare workers to explore more inventive ideas from social networks.

Social network analysis (SNA) enables analyzing the relevant information from the data stream, which provides the characterization and mapping of the hidden information during the intervention of the group of peoples in the social networks [12, 13]. Social network analysis mostly concentrates on the specifications of the interactions and interaction skills rather than focusing on the correspondence between the individual for establishing the effective interaction. Social network analysis is broadly utilized across a scope of regiments yet is frequently applied to enhance the adequacy and proficiency of dynamic cycles in business associations. Social network analysis renders great services in the dispersion research.

For the better advancements and the evolution of NIHR CLAHRC, York and Bradford utilized SNA to educate the turn of events and execute regarding custom-fitted conduct change mediations. These intercessions are pointed toward expanding the interpretation of exploration-based discoveries into nearby practice [14]. Nowadays, researchers concentrate on recognizing and supporting the relations to provide better take-up and the usage of the information [15].

The conventional methods provide a better perception of the victim experience with the aid of social media, which gathers the feedback of the patients from prominent web journals, such as Drugs.com and PatientsLikeMe.org. Moreover, information is also gathered from the negative impacts of the patients that are registered in the government web portals, such as FDA FAERS. Yet, the most advantageous methods utilize emerging technologies, such as Big Data analytics, to obtain more insights of the patients from the social media [16]. During the course of the recent years, web-based media, especially interpersonal interaction locales (SNSs), have been developed significantly. Such development, to a great extent, restrains all the existing obstructions in the social network for individuals to interface with each other, giving incredible potential to them, to keep up existing social ties and extend informal communities. Late investigations have exhibited that long range interpersonal communication capacities are powerful in improving clients' admittance to health-related data [17], interlocking the families in the lifestyle variation [18], and inspiring weight reduction [19]. The most astounding advancement in the web-based media coverage is the quick and ceaseless development of Facebook. Statistic reveals that in the US, approximately 65% of the web users utilize Facebook to refresh individual situations with companions or offer data [6]. Around the world, one in 7.7 individuals maintains the Facebook record, and near 530 million are day-by-day dynamic clients [20]. The advancements of the Facebook greatly influence the healthcare sector. Amidst US Facebook clients, 23% have monitored companions' very own well-being encounters or upgrades, 15% have recovered medical data on the site, and 9% have begun or joined a health-dependent gathering. Thus, Facebook holds an extraordinary potential to impact people's healthcare practices by molding their impression of normal practices and the assumptions that they set for themselves or by improving their admittance to an actually significant data. Late exploration has taken a gander at how clients use Facebook, as a nonexclusive SNS, for well-being data, the purpose of usage, and their view of the utilization [21, 22].

The existing healthcare analysis based on the social media extricates the health-concerned data from the social media, especially from the texture features [23, 24], or the interrelations between the structured features [25, 26]. As demonstrated in numerous investigations, in brain research and humanism [27], communications in informal organizations are significant elements to understanding practices of networks [28]. Web-based media, whenever used appropriately, can give significant bits of knowledge into understanding individuals' well-being practices at both individual and populace levels [29].

5.6 Conventional Strategies in Data Mining Techniques

A brief description of the existing system utilized for the opinion mining is elaborated in this section. The diagrammatic representation of the existing method is illustrated in Figure 5.2.

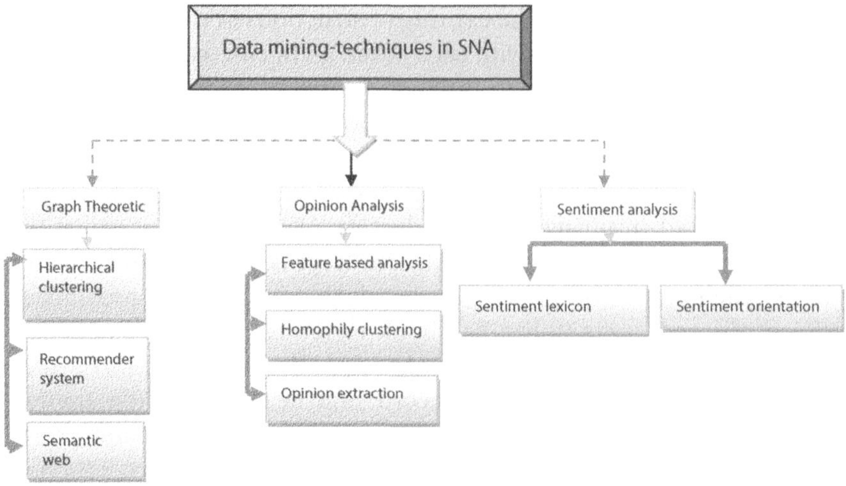

Figure 5.2 Diagrammatic representation of existing methods.

5.6.1 Graph Theoretic

Graph theory is one of the primary strategy utilized in the informal community investigation in the past statistics of the interpersonal organization idea. The methodology is applied in the SNA to decide significant highlights of the organization, such as the connections, and the nodes, such as the influencer and the followers. Influencers on informal community are perceived as customers that affect the daily activities or assessment of different customers via consumers or influence the options selected by different clients on the organization. Graph theory has winds up huge-scope data sets (like informal community information) because of the transmission capacity of the influencer to bypass the basic ideas of the visual characterization to run straightforwardly on information grids [30]. In [31], centrality measure was utilized to investigate the portrayal of force and impact that structures bunches and interrelations [32] among the interpersonal organization. The researcher of [33] utilized defined centrality metric way to deal with the study on the organizational framework and to categorize the node availability. Their work shaped an augmentation of a centrality approach that estimates the quantity of eased ways that subsists among the different nodes. The diagrammatic representation of Graph theory is illustrated in Figure 5.3.

a) Hierarchical clustering methods: A community is defined as the modest packed group gathering inside a bigger organization. Local area development is known to be one of the significant qualities of interpersonal organization destinations. Clients with comparable interest structure networks on interpersonal

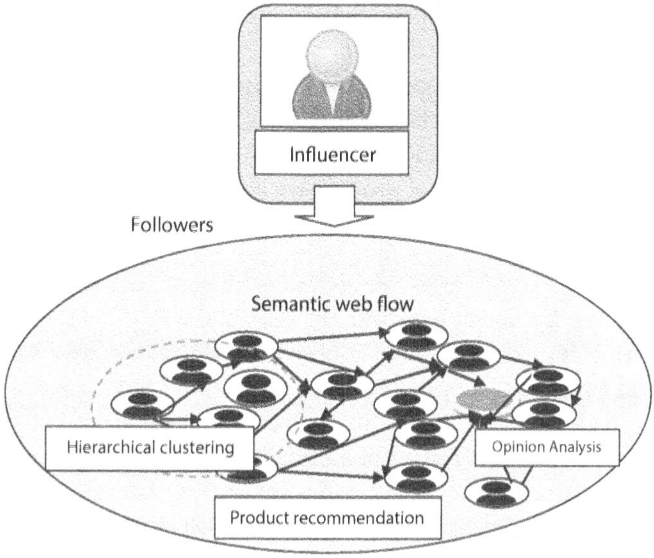

Figure 5.3 Diagrammatic representation of graph theory.

organization illustrate the solid sectional design. Networks on informal organizations, as related to some other networks in real-life, are intricate in their description and tremendous to recognize. Applying the suitable instruments in recognizing and comprehending the performance of organization networks is vital as this can be utilized to illustrate the effectiveness of the space they have in place. Various researchers have applied different clustering strategies to identify networks on interpersonal organization, with progressive grouping being generally utilized [34]. This method is the integration of numerous procedures used to bunch hubs in the organization to uncover strength of independent congregations, which is further utilized to disseminate the organization into networks. Vertex grouping has allocated with various leveled clustering strategies, diagram vertices can be resolved by adding it in a vector space so that pairwise distance between vertices can be estimated. Clients in a similar informal organization local area frequently prescribe things and administrations to each other, depending on their experience on particular things.

b) Recommender System in the social network community: Depending on the resemblance between hubs in informal community gatherings, collective filtering (CF) procedure, which assembles one of the three classes of the recommender framework (RS), can be utilized to deceive the relationship among clients [35]. Where CF's fundamental disadvantage is that of information sparsity, content-based (another RS strategy) investigates the constructions of the information to create suggestions. Nonetheless, the crossover approaches normally propose suggestions by consolidating CF and substance-based proposals. The demonstration in [36] presented a hybrid strategy named EntreeC, a framework that pools information-based RS and CF to suggest cafés. The work in [37] enhanced CF calculation by utilizing an eager execution of progressive agglomerative clustering to recommend approaching journals or the conference in which analysts, especially in the software engineering, can present their research.

c) Semantic web for Social network: The semantic web (SW) stage is utilized for the sharing and reutilization of conceivable information over various requisition and edges of the local area. Deep evaluation of SW improves quality of the information of SW community and conceives the integration of the SW. The work in [38] utilized Friend of a Friend (FOAF) to analyze how nearby and worldwide local area level gatherings generate and develop in enormous scope interpersonal organizations on the SW. The investigation reveals the advanced pattern of social designs and conjectures subsequent float. Similarly, application framework of SW-based social network evaluation frameworks provides the intellectual field media center of interpersonal organization evaluations connected with the ordinary layout of the SW to accomplish proficient recovery of Internet administrations. Besides, Voyeur Server [39] escalated the open-source Web-Harvest system for the assortment of online informal organization information to contemplate designs of privacy enhancement and of online logical amalgamation. Semantic web is a generally new territory in SNA and exploration in the field yet to be developed.

5.6.2 Opinion Evaluation in Social Network

As indicated by Technorati, around 75,000 new online journals and 1.2 million new posts provide the opinion on administrations, and the products are created each day [40]. Likewise, enormous information produced each moment on regular interpersonal organization locales are weighed down with assessment of clients with respect to assorted subjects going from individual to worldwide issues. The data mining techniques based on opinion analysis are elaborated in this section.

a) Feature-based analysis: The feature- or aspect-based opinion mining relies on the mining the most revived area of customers. It not only extricates the features from the reviews but also has to analyze the trustworthiness of the comments shared by the clients. Hence, the aspect-based analysis required to categorize the reviews as positive reviews and negative reviews.

b) Homophily clustering: The reviews shared by the clients in the social networks always reflect their own opinion; therefore, it fails to assure the general fact. The reviews will greatly influence the decision-making ability of the viewers. Hence, clustering technique is required for efficient data mining techniques to ensure the correct reviews of the products. The users, who share the same reviews or the opinions, are categorized under same class to form the cluster. This technique is known as homophily in the social organization.

c) Opinion extraction: The opinion extraction is the significant process to be accomplished in the data mining so as to sort out the real reviews about the product, persons, or the object. The most relevant information shared by the customers is extracted through the opinion extraction process.

5.6.3 Sentimental Analysis

The sentimental analysis receives huge attraction from the researcher after the publication of [41] and [42], in which the market sentiments are evaluated. The sentimental analysis is established in the data mining process so as to support the decision-making ability of the viewers.

a) Sentiment orientation: The widely utilized product attracts millions of the reviews from the clients, which may support the decision-making process. Meanwhile, the sellers promote their products with the aid of the sentiment analysis. To analyze the sentiments in the review, the hierarchical categorization techniques integrated with the machine learning techniques are presented in [43]. Hence, the flexible hierarchy is needed to attain the accurate classification.

b) Sentiment lexicon: The sentiment lexicon is characterized as the dictionary of the words that express the sentiments. The lexicon-based sentiments enhance the decision supporting system as it restricts the neutral reviews and through giving importance to the negative and positive comments.

5.7 Research Gaps in the Current Scenario

The research initiatives and conversations raised normal methodological issues, which are not rectified till now. To begin with, there requires the deep exploration to recognize the most significant health-related data from a healthcare organization. Also, there is no consensual strategy to decide the edge of shared patients that can be considered as a marker of genuine cooperation between medical services suppliers. Most of the research related to the medical data analysis utilizes an exact methodology by testing a few edges. Current methodological enhancements incorporate the limitation of linkages to medical care suppliers who share patients with each other during a scene of care: this makes it conceivable to restrict the possibly fake connections between medical services suppliers who treat similar patients for totally inconsequential conditions [44]. Besides, the investigation of medical services supplier networks raises some methodological issues. Most examinations just utilize a couple of organization measures among every one of those that are accessible (such as degree or thickness). This has additionally been noted in different fields of well-being research [45]. It is surely hard to track down measures that can satisfactorily describe coordination inside medical services supplier organizations. While those methodologies are utilized right now can help portray various models of healthcare coordination, it stays hard to make determinations on the nature of coordination utilizing such measures. There needs to be a solid framework for connecting network measures with patient results to distinguish the best models of coordination. Also, the change of bipartite organizations into unipartite organizations may require a more grounded fundamental framework on the portrayal of bipartite organizations. Third, research applying SNA to medical information needs to consider the restrictions innate to the utilization of such information. All the time, these just incorporate medical services experiences and do not give any data on well-being and social consideration. Also, these data sets ordinarily do not make it conceivable to distinguish the individual healthcare experts working in emergency clinics, which oblige the conduction of staggered investigations. Furthermore, while the utilization of shared patients records to recognize medical care supplier networks has been approved [44], it very well may be valuable to supplement it with different markers, like the presence of composed or email correspondences between healthcare experts, which are, for the most part, not revealed in guarantee information. It seems important to match full-scale quantitative methodologies with more sociological or related methodologies, for example, the investigations introduced in the last three commitments of the workshop, to guarantee an exhaustive comprehension of the systems at play (for example, whether an organization is based on doctor or patient practices). These methodological issues in the use of SNA to medical care information support the advancement of solid cooperative energies between research groups utilizing these techniques as a component of a multi-disciplinary methodology, which requires an exchange of both the quantitative and subjective methodologies between the healthcare experts.

5.8 Conclusion and Challenges

The challenges and the research issues encountered in the analysis of the social network using the data mining techniques are identified and enlisted as follows:

- Structural- or linkage-based analysis: The structural-based analysis is the exploration of the linkage conduct of the informal organization to learn assessable nodes, connections, networks, and approaching spaces of the organization [46].
- Dynamic and the static analysis: Some of the static evaluation, such as the bibliographic organizations, are ventured to be simpler to execute in the streaming organization. Yet, in static examination, it is assumed that interpersonal organization varies in accordance to the variation with respect to the time, and investigation on the whole organization should be possible in group mode. Hence, it is a perplex process to analyze the social networks, such as YouTube and the Facebook. Gathering information from the social network is a tedious process as the data on these organizations are produced fast and with limits. Dynamic evaluation of these organizations is regularly implemented in the space of connections between elements [47], transient occasions on interpersonal organizations [6], and developing networks [48]. Having introduced a portion of the examination issues and difficulties in SNA, the accompanying areas and subsegments present the outline of various information mining approaches utilized in breaking down interpersonal organization information [6].

Regardless of methodological difficulties, the utilization of SNA to medical care information empowers us to address new evaluation inquiries in the field of healthcare administration research. It supplements regular methodologies like planning, or hypothetical systems like dispersion speculations, through unmistakable, probabilistic, or displaying strategies. It likewise opens up research points of view that have not been investigated a lot to date, including the investigation of advancements after some time by getting determinants and components of progress inside medical services supplier organizations. Moreover, the use of SNA to medical services information gives freedom to help the dynamics in a setting where joint efforts between healthcare experts are at the focal point of coordination issues, with suggestions for the development of new subsidizing instruments. While the immediate interpretation of exploration projects on medical care supplier networks into proof-based dynamic has been negligible, because of the somewhat late advancement of such undertakings, the organizations recognized by Thérèse Stukel and her associates [49, 50] have propelled the production of incorporated frameworks to improve care for high-need, significant expense patients in Ontario. The distinguishing proof of previous casual medical services supplier organizations can in fact shape a judicious reason for growing more conventional organizations, like ACOs in the US, or to screen care execution without restricting the attribution of duty to a solitary supplier. Notwithstanding, such an exchange from exploration to dynamic requires further work on the most ideal ways that healthcare specialists and related associations may utilize the devices created by analysts to improve coordination. Moreover, scientists ought to guarantee that their techniques are clear as SNA, particularly when it utilizes the latest modeling apparatuses, like exponential graph models, which can be unpredictable. The utilization of perception instruments can be a significant method of imparting discoveries to strategy producers yet their logical adequacy (specifically for the translucent based graph theory) ought to be validated.

References

1. Matengo W., Otsieno E., Wanjiru K., Big Data Analytics in Healthcare. In: *Digital Health in Focus of Predictive, Preventive and Personalised Medicine*, Chaari L. (eds) Advances in Predictive, Preventive and Personalised Medicine, vol. 12, Springer, Cham, Springer Nature Switzerland AG, 2020. https://doi.org/10.1007/978-3-030-49815-3_15
2. Alotaibi S., Mehmood R., Katib I. The Role of Big Data and Twitter Data Analytics in Healthcare Supply Chain Management. In: *Smart Infrastructure and Applications*, Mehmood R., See S.,Katib I., Chlamtac I. (eds), EAI/Springer Innovations in Communication and Computing, Springer, Cham, Springer Nature Switzerland AG, 2020. https://doi.org/10.1007/978-3-030-13705-2_11
3. The State of Social Intelligence 2019: Painting a Masterpiece with Social Data. Demographics Pro Inc (White Paper), 2019. https://www.demographicspro.com/
4. Naqishbandi T.A., Ayyanathan N. Clinical Big Data Predictive Analytics Transforming Healthcare - An Integrated Framework for Promise Towards Value Based Healthcare. In: *Advances in Decision Sciences, Image*

Processing, Security and Computer Vision, Satapathy S., Raju K., Shyamala K., Krishna D., Favorskaya M. (eds). Learning and Analytics in Intelligent Systems, vol 4, Springer, Cham, Springer Nature Switzerland AG, 2020. https://doi.org/10.1007/978-3-030-24318-0_64

5. Weeks R., *Exceptional Examples of Social Media Marketing in Healthcare*, LinkedIn Pulse, 2015. https://www.linkedin.com/pulse/8-exceptional-examples-social-media-marketing-healthcare-rachel-weeks
6. Stahl, F., Gaber, M. M., & Adedoyin-Olowe, M., A survey of data mining techniques for social media analysis. *Journal of Data Mining & Digital Humanities*, France, 2014.
7. Smith, K.T., Hospital Marketing and Communications via Social Media. *Serv. Market. Q.*, 38, 3, 187–201, 2017.
8. Khan, G.F., Swar, B., Lee, S.K., Social Media Risks and Benefits: A Public Sector Perspective. *Soc Sci. Comput. Rev.*, 32, 5, 606–627, 2014.
9. Fan, W. and Gordon, M.D., The Power of Social Media Analytics. *Commun. ACM*, 57, 6, 74–81, 2014.
10. Baktha K., Dev M., Gupta H., Agarwal A., Balamurugan B., Social Network Analysis in Healthcare. In: *Internet of Things and Big Data Technologies for Next Generation Healthcare,* Bhatt C., Dey N., Ashour A. (eds). Studies in Big Data, vol. 23, Springer, Cham, Springer International Publishing AG, 2017. https://doi.org/10.1007/978-3-319-49736-5_13
11. Sanson-Fisher, R.W., Diffusion of Innovation Theory for Clinical Change. *Med. J. Aust.*, 180, S55–S56, 2004.
12. Cross, R.L., Cross, R.L., Parker, A., ·e *Hidden Power of Social Networks: Understanding How Work Really Gets Done in Organizations*, Harvard Business Press, Massachusetts, 2004.
13. Hanneman, R. A. and Riddle, M., *Introduction to social network methods*, University of California, Riverside, CA, 2005.
14. Hanbury, A., Thompson, C., Wilson, P.M., Farley, K., Chambers, D., Warren, E., Bibby, J., Mannion, R., Watt, I.S., Gilbody, S., Translating research into practice in Leeds and Bradford (TRiPLaB): A protocol for a programme of research. *Implement. Sci.*, 5, 1, 1–6, 2010.
15. Chambers, D., Wilson, P., Thompson, C., Harden, M., Social Network Analysis in Healthcare Settings: A Systematic Scoping Review. *PloS One*, 7, 8, e41911, 2012.
16. DiMatteo, M.R., Social Support and Patient Adherence to Medical Treatment: A Meta-Analysis. *Health Psychol.*, 23, 2, 207, 2004.
17. Freyne, J., Berkovsky, S., Kimani, S., Baghaei, N., Brindal, E., *Improving Health Information Access Through Social Networking*, 334–339, 2010.
18. Baghaei, N., Kimani, S., Freyne, J., Brindal, E., Berkovsky, S., Smith, G., Engaging Families in Lifestyle Changes Through Social Networking. *Int. J. Hum. Comput. Interact.*, 27, 10, 971–990, 2011.

19. Hwang, K.O., Ottenbacher, A.J., Green, A.P., Cannon-Diehl, M.R., Richardson, O., Bernstam, E.V., Thomas, E.J., Social Support in An Internet Weight Loss Community. *Int. J. Med. Inform.*, 79, 1, 5–13, 2010.
20. Zhang, Y., He, D., Sang, Y., Facebook As A Platform For Health Information and Communication: A Case Study of a Diabetes Group. *J. Med. Syst.*, 37, 1–12, 2013.
21. Newman, M.W., Lauterbach, D., Munson, S.A., Resnick, P., Morris, M.E., It's not that I don't have problems, I'm just not putting them on Facebook: challenges and opportunities in using online social networks for health, in: *Proceedings of the ACM 2011 Conference on Computer Supported Cooperative Work*, pp. 341–350, 2011.
22. Zhang, Y., He, D., Sang, Y., Facebook as a platform for health information and communication: a case study of a diabetes group. *J. Med. Syst.*, 37, 3, 1–12, 2013.
23. Culotta, A., Towards detecting influenza epidemics by analyzing Twitter messages, in: *Proceedings of the First Workshop on Social Media Analytics*, pp. 115–122, 2010.
24. Culotta, A., Estimating county health statistics with twitter, in: *Proceedings of the SIGCHI Conference on Human Factors in Computing Systems*, pp. 1335–1344, 2014.
25. Nguyen, T., Larsen, M., O'Dea, B., Nguyen, H., Nguyen, D. T., Yearwood, J., Christensen, H., Using spatiotemporal distribution of geocoded Twitter data to predict US county-level health indices. *Future Generation Computer Systems*, 110, 620–628, 2020.
26. Nguyen, T., Nguyen, D.T., Larsen, M.E., O'Dea, B., Yearwood, J., Phung, D., Venkatesh, S., Christensen, H., Prediction of population health indices from social media using kernel-based textual and temporal features, in: *Proceedings of the 26th International Conference on World Wide Web Companion*, pp. 99–107, 2017.
27. Wilson, C., Boe, B., Sala, A., Puttaswamy, K.P., Zhao, B.Y., User interactions in social networks and their implications, in: *Proceedings of the 4th ACM European Conference on Computer Systems*, pp. 205–218, 2009.
28. Akbari, M., Relia, K., Elghafari, A., Chunara, R., From the user to the medium: neural profiling across web communities, in: *Proceedings of the International AAAI Conference on Web and Social Media*, vol. 12, 2018.
29. Nguyen, H., Nguyen, T., Nguyen, D.T., A graph-based approach for population health analysis using geo-tagged tweets. *Multimed. Tools Appl.*, 80, 5, 7187–7204, 2021.
30. Burt, R.S., Brokerage and closure: An Introduction to Social Capital, Oxford University Press, UK, 2005.
31. Burt, R.S., *Brokerage and closure: An Introduction to Social Capital*, Oxford University Press, UK, 2005.

32. Bethard, S., Yu, H., Thornton, A., Hatzivassiloglou, V., Jurafsky, D., Automatic Extraction of Opinion Propositions and Their Holders, in: *Proceedings of AAAI Spring Symposium on Exploring Attitude and Affect in Text*, p. 2224, 2004.
33. Ghosh, R. and Lerman, K., Parameterized centrality metric for network analysis. *Phys. Rev.*, 83, 6, 066118, 2011.
34. Newman, M.E.J., Random graphs as models of networks, in: *Handbook of Graphs and Networks*, pp. 35–68, 2003.
35. Liu, F. and Lee, H.J., Use of social network information to enhance collaborative filtering performance. *Exp. Syst. Appl.*, 37, 7, 4772–4778, 2010.
36. Burke, R., Hybrid recommender systems: Survey and experiments. *User Model User-adapt. Interact.*, 12, 4, 331–370, 2002.
37. Pham, Manh Cuong, Yiwei Cao, Ralf Klamma, and Matthias Jarke. A clustering approach for collaborative filtering recommendation using social network analysis. *J. Univers. Comput. Sci.* vol. 17, no. 4, pp. 583–604, 2011.
38. Zhou, L., Ding, L., Finin, T., How is the Semantic Web evolving? A dynamic social network perspective. *Comput. Hum. Behav.*, 27, 4, 1294–1302, 2011.
39. Murthy, D., Gross, A., Takata, A., Bond, S., Evaluation and development of data mining tools for social network analysis, in: *Mining Social Networks and Security Informatics*, pp. 183–202, 2013.
40. Kim, P. *The forrester wave: Brand monitoring, Q3 2006. Forrester Wave (white paper)*, 2006.
41. Das, S. and Chen, M., Yahoo! for Amazon: Extracting market sentiment from stock message boards, in: *Proceedings of the Asia Pacific Finance Association Annual Conference (APFA)*, vol. 35, p. 43, 2001.
42. Tong, R.M., An operational system for detecting and tracking opinions in on-line discussion, in: *Working Notes of the ACM SIGIR 2001 Workshop on Operational Text Classification*, vol. 1, 2001.
43. Keshtkar, F. and Inkpen, D., Using sentiment orientation features for mood classification in blogs, in: *Proceedings of International Conference on Natural Language Processing and Knowledge Engineering*, pp. 1–6, 2009.
44. Landon, B.E., Keating, N.L., Onnela, J.P., Zaslavsky, A.M., Christakis, N.A., O'Malley, A.J., Patient-sharing networks of physicians and healthcare utilization and spending among Medicare beneficiaries. *JAMA Int. Med.*, 178, 1, 66–73, 2018.
45. Shelton, R.C., Lee, M., Brotzman, L.E., Crookes, D.M., Jandorf, L., Erwin, D., Gage-Bouchard, E.A., Use of social network analysis in the development, dissemination, implementation, and sustainability of health behavior interventions for adults: A systematic review. *Soc Sci. Med.*, 220, 81–101, 2019.
46. Aggarwal, C.C., An introduction to social network data analytics, in: *Social Network Data Analytics*, pp. 1–15, 2012.
47. Papadopoulos, S., Kompatsiaris, Y., Vakali, A., Spyridonos, P., Community detection in social media. *Data Min. Knowl. Discovery*, 24, 3, 515–554, 2012.
48. Fortunato, S., Community detection in graphs. *Phys. Rep.*, 486, 3-5, 75–174, 2010.

49. Stukel, T.A., Glazier, R.H., Schultz, S.E., Guan, J., Zagorski, B.M., Gozdyra, P., Henry, D.A., Multispecialty physician networks in Ontario. *Open Med.*, 7, 2, e40, 2013.
50. Galety, M.G., Data security in big data using parallel data generalization algorithm. I*nternational Journal of Advanced Trends in Computer Science and Engineering*, 8, 1, 75–79, 2019.

6

Pragmatic Analysis of Social Web Components on Semantic Web Mining

Sasmita Pani[1], Bibhuprasad Sahu[1*], Jibitesh Mishra[2], Sachi Nandan Mohanty[3] and Amrutanshu Panigrahi[4]

[1]*Department of Computer Science and Engineering, Gandhi Institute for Technology, Bhubaneswar, Odisha, India*
[2]*Department of Master in Science and Computer Application, College of Engineering, Bhubaneswar, Odisha, India*
[3]*Department of Computer Science and Engineering, College of Engineering, Pune, India*
[4]*Soa Deemed to be University, Bhubaneswar, Odisha, India*

Abstract
To obtain information in a meaningful manner, it is necessary to analyze and design them semantically. Otherwise, information cannot be retrieved semantically. In this study, we proposed an integrated ontology model for providing semantic information to farmers, as well as vendors in the agricultural sector. The agriculturalal system is taken here as a domain upon which semantic web is built, and agriculture information system is considered from the user's perspective, and information is analyzed using ontology. Here the farmers relate to the social web application when they search for information on mobile devices. We adopted the questionnaire method for collecting primary data. Twenty farmers having a mean age of 29 years and 20 agricultural vendors, with a mean age of 35 years, those providing fertilizers and machine equipment to the agriculture sector, participated in this study. All the participants in this study were men and were from the eastern region of Odisha, India. The domain ontology has been built based on agricultural and social ontology consisting of social parameters or social web components, such as wikis, podcasting, and social networks. In the second step, an analysis is done about the usability of different social web components on the domain ontology. Also, in the third step, combined ontology is built by taking domain ontology and social ontology. The class axioms

**Corresponding author*: bibhuprasad@gift.edu.in

Mohammad Gouse Galety, Chiai Al Atroshi, Bunil Kumar Balabantaray and Sachi Nandan Mohanty (eds.) Social Network Analysis: Theory and Applications, (83–108) © 2022 Scrivener Publishing LLC

and property axioms have been defined in the combined ontology based on the analysis of social web components across an agricultural domain. This study proposes an integrated model approach with social semantic ontology under a specific (agricultural) domain, which is comprised of domain ontology and social ontology. This integrated approach is used for establishing social semantic ontology. The result reveals that social networking, content hosting, and blogs are useful in providing information for the demand of the users.

***Keywords*:** Social ontology, FOAF, domain ontology, class axioms, property axiom

6.1 Introduction

The web application includes different social applications, such as wikis, podcasting, blogs, content hosting, social networking, e-portfolios, and social bookmarking [1, 2]. Social web components form the network between users from different organizations, and it helps to construct bridges among the people. Social networking sites construct a variety of users with whom the users share the connection among themselves. Social networking sites construct a public or semipublic profile where the users can see and visit a list of connections made by the users within the system. The social web components areas (www.unimelb.edu.au), social networking, blogs, and pod casting [3], where the user can be connected with a group of users, find solutions to their queries and obtain information. Social web applications drive the relationship among the social web components and users to usefulness to fulfil the requirements of the user that satisfy the mobile web applications. A social network is used to construct web applications that provide users to create their profiles so that they can be connected with other users within the network [4], irrespective of any domain. The social network has made a drastic change and introduced social connections among users in web 2.0. It is provided by the use of meaningful information extraction and exchanging user-generated content. A social networking web application such as Facebook focuses on finding old friends and making new friends. LinkedIn is a social networking web application, or a professional networking site that focuses on professional networking among professionals. A wiki is a social application that collects and organizes content, created and revised by its users. In Wikipedia, the users easily access the information. Wikis drive the way to grow knowledge about a particular subject area. It also provides the best strategies in a given field or how to use a specific piece of software. Wikis develop a community of users. This social application includes the original contributor of an article and others who have done changes and made revisions. Blogging is a social web application that produces content to post in the form of blogs and establishes social relationships among users. It also provides commentary information subject to any context ranging from sports to politics. Podcasting refers to the creation and download of data in the form of audio, video, and PDF files through the web. Podcasting provides great support in e-learning and m-learning application. It also provides information related to agriculture, the healthcare system, education, academia, and so on. A content hosting social application is a web hosting application that specifically hosts the user's created data on social networking web applications. For example, content hosting in the curriculum provides sharing of curriculum-related resources. It also provides learners to retrieve the user-generated content for doing their specific applications. The structure of social networks and their strategic positions are identified by social network analysis (SNA), which includes graph-based [17] algorithms.

Online social network applications, like Facebook, are used to construct social networks where people interconnect with each other and exchange data on the web. The SNA operator uses a semantic web that makes the involvement of graph-based representations. Those graph-based representations are also used when analyzing social networks and their relations and interactions. The directed labeled graph structure of the resource description framework (RDF) is suitable for representing social knowledge and its produced metadata. The diversity of interactions and relationships associated with parameterized SNA metrics is analyzed and handled through SNA. Using data mining techniques in SNA could be identified as (a) linkage-based and structural-based analyses, which specifically provide the interconnection between social networks to achieve the relevant nodes, links, communities, and imminent nodes in the social network. (b) Dynamic analysis and static analysis, such as in bibliographic networks, are found to be the easiest way for providing analysis in streaming networks. In static analysis, it is assumed that social network changes gradually from time to time, and the analysis is done for the entire network through cluster processing. However, dynamic analysis of streaming networks, like Facebook and Youtube, is found to be very difficult to carry out. Data on these networks are produced at high speed and capacity.

Some data mining methods and algorithms are to describe and analyze the social network. These methods are based on similarity measurement algorithms and inductive logic programming (ILP), which are useful here to analyze social networks. ILP [18] is a field that spans along with machine learning and logic programming, mainly concerned with finding new knowledge from the data. Analyzing social networks is a major application of relational data. The advancement of relational data mining gives a more powerful tool for SNA. Similarly, measures [18] are very effective in link prediction, which determines whether there is any link between the two actors or not. In the social network *G*, the similarity measure functions for each pair of nodes: <*x*, *y*>, is given as a link score *(x, y)*. In some applications, the function can be seen as the topology structure of network G and it calculates the degree of similarity for each node x and y. Social network analysis is a set of norms and methods to interpret the structure of the social network and its properties, which is also called structural analysis. It mainly focuses on the structures and attributes of a social relation constituted by different social organizations.

For analyzing social networks, many techniques are used, such as formal methods, graphic display matrices, and statistical models. Many formal methods [19] are the combination of mathematics and graphs, which are used for representing social network. It is because they deal with graph processing and rule-based techniques, which are used for retrieving data in the social network. These techniques described the data in the social network more compactly and systematically. A graphic display technique that consists of nodes and edges is used for representing social relationships among people. Matrices are also used for this SNA. The statistical models provide modeling on social networks. It assumes that there are n entities called actors and information as binary relations between them. The binary relations are represented as an order of $n \times n$ matrix say Y, where entities are usually represented as nodes and the relations as arrows between the nodes. The value of Y_{ij} is 1 if actor i *is* related or connected to j and 0 otherwise. For example, if i is a friend of j, then $Y_{ij} = 1$.

Ontology specifies the meaning of annotations and also provides a vocabulary of terms. By using it, new terms can be formed by combing existing ones. In the web application, the meaning of such terms can be formally specified through it. In this study, agriculture systems have been taken as a domain on which semantic web is built and agriculture information system is taken into account from the user's perspective, and information is analyzed using ontology. Ontology provides a well-organized and defined mechanism for the modeling, querying, and retrieving of the required information. Ontology emphasizes sharing a common understanding of the structure of information among people or users, irrespective of any domain. It also enables a user to reuse and analyze domain knowledge. Ontology is the key technology for constructing, organizing, and exploiting information for the efficient retrieval of knowledge [5]. The study of ontology and its usage is similar to one of the fields in artificial intelligence.

The ontology forms the cornerstone for the semantic web, and it can be used in e-commerce and various application fields, such as e-science, digital libraries, bioinformatics, and medicine. Ontology is a collection of conceptions that falls on any domain of discourse where properties of each concept are described by various features and attributes. It also describes the restrictions that are associated with each property of conceptions. By enabling the class axioms and property axioms irrespective of any domain, information retrieval can be done in a meaningful and semantic manner. A semantic web application is built using ontology. The semantic web is an extension of the current web in which information is constructed and organized in a meaningful and well-defined manner. The semantic web is an approach for constructing the web as an intelligent and intuitive search engine where information can be retrieved semantically.

In the first step, the domain ontology is built based on agriculture and social aspects consisting of social parameters or social web components, such as wikis, podcasting, and social networks. In the second step, an analysis is done about the usability of different social web components on the domain ontology. And in the third step, we have built combined ontology by taking both domain ontology and social ontology. We have established the class axioms and property axioms in the combined ontology based on the analysis of social web components across the agriculture domain.

6.2 Background

6.2.1 Web

The worldwide web (www) is the largest information provider, and the organizer had much success since its advent. The www (commonly known as the web) is not synonymous with the Internet but is the most prominent part of the Internet that can be defined as a technology-based social system to interact with humans, based on technological networks. Web 1.0 is the first generation of the web, which is an information provider to provide information to people. The early web provided content based on search and limited user functionality. Web 2.0 was a read–write web that provides extended user functionality, content generation, and publishing content on the web. It also provides social interactions among the users, whereas Web 3.0 establishes semantics between data on the web. Web 3.0 combines semantics with the features of Web 2.0.

Web 3.0 is a semantic web, which is the integration of contextual search, personalized search, and deductive reasoning. A semantic web application is built using ontology. Ontology specifies the meaning of annotations and also provides a vocabulary of terms. By using it, new terms can be formed by combing existing ones. In the web application, the meaning of such terms is formally specified through it. It embeds intelligence on the web and intends to decrease human tasks and decisions and leave them to machines by providing machine-readable contents on the web [6]. The users belonging to any domain can find domain-specific information from the web and connects with people in the near world through social web components, like social networks, podcasting, blogs, e-portfolios, and social bookmarking. The social web component accomplishes a set of people connected by others through friendship, co-working, and information exchange. Web 4.0 [30] is a smart ultraintelligent system with a camera and face recognition system. Web 4.0 is a mobile web that provides information retrieval over the semantic web, which is a ubiquitous and pervasive system. Hence, to support web 4.0, the mobile device should be embedded with efficient mobility functionality. Accessing the web on mobile devices focuses on the four aspects of mobile devices, such as device aspect, user aspect, mobility aspect, and social aspect of social web component [7].

6.2.2 Agriculture Information Systems

The advancement in the technologies of wireless communications has brought opportunities for different mobile applications running on ubiquitous and pervasive-based systems. Through enabling technologies, mobile application users may access services, such as payment, booking, and so on, through mobile devices and find information anywhere and anytime. The government has started the e-choupal project, which deals with the establishment of Internet centers in rural areas, such as villages, where a farmer can find and access agriculture-related information more efficiently. The e-choupal portal [8] provides data on weather, new advanced farm techniques, risk handling, and knowledge and purchases of high-quality agricultural products in local languages to the farmers. The e-choupal system has obtained India tobacco company (ITC) to take necessary action for organizing and managing its own IT network in rural India. This system also enables to identify and give training to a local farmer for managing the system itself.

Using e-choupal, the farmers can know closing prices on local "mandis," as well as track global price trends on a regular basis. The farmer and vendor can obtain information on new advanced farming machinery and technology. The farmers also make use of the e-choupal system to place an order on agricultural-related products and other products at prices lesser than what are available from village traders. E-choupal [31] began to start providing third-party services in rural India. Currently, there are more than 160 partners from domains that are providing information on seeds, crops, and biocides. They are also assisting farmers on finance, insurance, and employment for selling their products to rural consumers through e-choupal's channel. In 2004, the government has established Kisan call centers everywhere in India to deliver information about agriculture to the farmers in their local languages.

M-Krishi is a high-end technical service organized by Tata Consultancy Services (TCS) in 2007 in India for giving personalized information to farmers based on producing crops, market, and weather forecast. M-Krishi also involves different kinds of sensors to be fixed in the areas for extracting humidity of soil and weather types. The current M-Krishi established an automatic weather station, which is deployed at the midpoint of the village, that provides information on soil/weather discrimination for the area. The information related to crop, soil, and micro-environment is gathered by sensors and sent to a central server using the Internet. M-Krishi is based on a computerized database that can provide quick responses to a farmer because many of them are generic. Questions that are more specific or sophisticated are sent to experts through the Internet. The farmers can see photographs and information regarding soil using sensors through messaging, and their response is given via SMS to the farmer. Farmers receive responses to their questions based on agriculture through the proper channel within 24 hours.

The joint venture IKSL (Indian Farmer Fertilizer Cooperative Ltd [IFFCO] Kisan Sanchar Limited) is established by the association of mobile operator Bharti Airtel and IFFCO [10] in 2007. This company furnishes information on prices of the market, agricultural machinery, product availability, weather updates, and so on. Indian Farmer Fertilizer Cooperative Ltd brings five free daily voice-based messages in their respective local languages except for Sundays. The International Institute of Information Technology (IIIT), Hyderabad India, had formed a tele-agriculture project named e-sagu [9, 11, 26] in association with Media Lab, Asia in the year 2004. E-sagu provides agricultural-related advice once a week starting from different phases. This service enhances agriculture productivity and brings down the farming-related cost. It also increases the quality of agricultural products. The organizer gathers agriculture-related data including soil, water resources, farming products, finance, and so on, and sends these to the e-sagu system. Every week, the administrator or organizer assembles the data on farm operations from the previous week. He then makes his observation on the various problems regarding farming and takes some digital photographs. He sends this information, including photographs through a disk, and is sent through courier. The agricultural experts from different departments analyze the problems and develop agriculture-related advice. The advice is downloaded through the Internet from the e-sagu system. This way, each farmer gets a response in each phase of the agricultural system and can be benefited.

6.2.3 Ontology in Web or Mobile Web

Ontology is a widely accepted, efficient, and popular mechanism for data modeling in web-enabled domains [25]. The rising semantic web technologies and mobile communication fields get assembled to achieve the convergence formed by mobile and web-based systems toward the vision of ontology-enabled ubiquitous pervasive-based systems [12]. Semantic and ontology technologies are being used to make advancements in the seamless integration of web-based and pervasive systems. Ontology models classes and constructs relationships in a semantic manner. Ontology-driven applications help users to find information based on inquiry effectively [13] because they contain properties, such as expressions, extensions, ease of sharing data, and logic reasoning support. However, because the mobile world has been associated with the web world in delivering new value-added services, thus semantics in ubiquitous mobile communication possess greater importance and potential.

CACOnt [14] is an ontology-based model for modeling generic context ontology by capturing contexts. This ontology-based general and extensible context model hierarchically includes general and domain-specific ontology and provides a hybrid approach of context reasoning based on ontology and rules. COMET [15] provides a semantically descriptive model for mobile learning, which divides the context into three parts, such as activity context, learner-centric context, and environmental contexts, and models these contexts using ontology. The initial prototype of the GCoMM, a generic context model [16], is a multi-domain context-aware platform where context reasoning is done by parsing and interfacing the mechanism of rules and context objects to provide context-aware services.

6.3 Proposed Model

The different aspects of social networks or social web components can be analyzed through descriptive methods and statistical methods. Also, the social network semantic model (FOAF) [19] is built through ontology

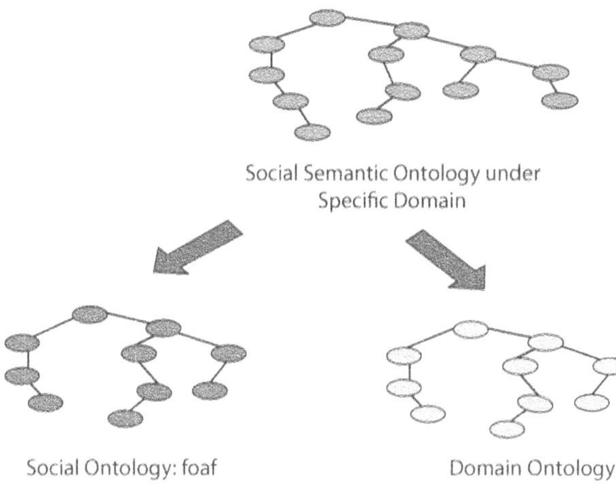

Figure 6.1 Building social semantic ontology for specific domain.

using RDF schema. Because these social web components produce a larger impact on people related to any domain, hence, it is necessary to analyze these social web components on the semantic web specific to any domain. We will analyze how social networks, like Facebook, Twitter, and LinkedIn, work on the semantic web. These analyses are done over the semantic web under any domain through generating questionnaires, Multiple Choice Questions, and find their response. Domain ontology [13] is built on an agriculture system, e-learning system, healthcare system, and the users such as farmers, students, or doctors, can connect themselves and exchange information on the web. We proposed an integrated ontology model that covers social semantic ontology and domain ontology (Figure 6.1).

6.3.1 Developing Domain Ontology

The motivation toward the study is to develop a process or technical architecture to build ontology for the agricultural domain. In section 6.2.2, various e-agricultural information systems are discussed, which provide information about soil, crop, weather, and farming techniques, and so on. These e-agricultural systems dealt with enormous data, but the data are not designed in a structured way and are also not represented in a meaningful or semantic manner. These web-based information systems are not specifying vocabulary about the terms and do not have formal meaningfulness of the terms. Hence, this web-based information system is not handling data consistency and structured and semantic data. Ontologies are used in web-based applications to overwhelm these things and build semantic data for any domain. Ontology is used on the web and provides meaningful annotations and vocabulary of terms about a certain domain. To build an ontology for any domain, it is necessary to analyze the information required for a user and build an ontology for that domain on the web through web ontology language (OWL).

We visited the Agricultural Promotion & Investment Corporation (APICOL) of Odisha, India, and found out the different phases [20], which include the following: (a) crop identification phase; (b) soil preparation and sowing phase; (c) crop growing and protection phase; (d) harvesting phase; (e) storage, distribution, and selling phase.

In the crop identification phase, a particular crop is selected, depending on the soil, weather, and zone conditions. The correct identification of the crop can lead to a successful harvest and productivity. The crop identification phase leads to the seed collection from different sources, like agriculture seed banks, previously harvested stock, and so on. Proper crop identification reduces the demerits of the farming procedure involved to produce a certain crop. The soil plays an important role in the crop identification phase as the crop to be selected has to comply with the soil properties supporting its growth. The weather affects the particular crop in its growth and productivity. An incompatible weather condition adversely affects crop productivity and can lead to crop loss. Hence, suitable weather condition governs the crop selection. According to the soil and weather conditions, a particular agricultural zone is determined. Hence, the agricultural map of India is segregated into various zones, like Gangetic plains, Deccan zone, coastal zone, desert zone, and Himalayan zone [21]. The weather of a particular zone helps the farmer to choose a particular crop suited for a season.

The farmer or user interaction with the desktop/pervasive systems enables various agriculture information systems to be developed, which have already been discussed in section 6.2.2. The user context includes the information needed by a user according to its role. Here, the user (farmer) needs information about identifying proper crop, soil, weather, season, and zone through a mobile web application. These information elements are described in the agricultural ontology or deriving ontology. Further, those information elements are designed with OWL descriptive logic (OWL-DL) in mobile web applications to provide semantic information to the farmers.

In the soil preparation and sowing phase, the pre-sowing procedure is followed where the land bed is prepared according to the particular crop which is selected. The soil is cleaned and processed by adding fertilizers and others biocides. Then, the sowing process is followed like the broadcasting and the transplantation method, depending on seed types, like high yield and low yield seeds. The soil preparation and sowing process make efficient use of modern farming equipment, which increases the efficiency of this process. In this phase, the user (farmer) needs information about the soil preparation and sowing of the particular crop for cultivating through mobile web applications. Hence, the information elements include soil type, sowing methods, and farming equipment, which are described in the agriculture ontology.

In this crop growing and protection phase, the crop growing process is done. During the growing phase (State of Indian Agriculture, 2013) [22], proper care and protection are taken to enhance the efficiency of growth in time. Proper care and protection lead to defect-free harvest and good crop quality, thus increasing its productivity and profit. Crop growth and protection are done using efficient irrigation techniques, fertilizers, biocides usage, and farming equipment. The information elements include all the things described above which are further discussed in the derived ontology. In the harvesting phase, the crop is harvested to be processed for packing, storage, and distribution. During harvesting, several harvest methods are applied for quick and efficient harvesting to generate a clean and quality product. These methods can be either manual or mechanical by the application of farming equipment, which can be either traditional or modern harvesting tools. Hence, both the harvesting techniques and the farming equipment are essential for a farmer, and the farmer needs information about the process and techniques of equipment through the mobile web applications. These information elements are shown in the derived ontology.

The final and last phase is the storage, distribution, and selling phase, which is called a post-harvest phase. In this phase, the crop is packed efficiently for storage in clean and hygienic environments for distribution and selling. The crops are packed using scientific techniques in cloth, jute, or polythene bags and are stored in natural or cold storage rooms. The crop is distributed for either commercial purposes or through public distribution system for government-sponsored schemes [23, 27, 28]. The public distribution system is a government-sponsored fair price shop to distribute the food grains to the underprivileged section of the society. Commercial distribution is the market selling off food grains and other crops. The pricing is also determined based on the above distribution criteria [24]. The prices can be segregated into minimum support price (MSP), which is the PDS price, wholesale price, and retail price. Hence, the user, which is the farmer, needs information about storage, packing, distribution, and pricing through the mobile web application. Further, these information elements are described in the agriculture ontology.

6.3.2 Building the Agriculture Ontology with OWL-DL

OWL ontology describes classes about a particular domain and represents relationships among the classes. It builds various classes, such as crop, soil, fertilizers, farming equipment, and so on, and describes relationships/properties by using OWL. We define class axioms about the classes and build class hierarchy by using OWL description logic, which is a sublanguage of OWL and uses description logic reasoner to check the consistency in the agriculture ontology and compute the class hierarchy. The property restrictions and object property characteristics are also described for a property or relationships that hold among the classes by using OWL-DL in protégé 5.0 beta for agriculture ontology to provide semantic information to farmers (users) through the web application.

6.3.2.1 Building Class Axioms

In agriculture ontology, the classes, such as agriculture system, crop, farmer, fertilizers, pesticides, farming equipment, seed, and so on, are taken. All these classes are made as disjoint class so that simultaneously the instance of one class cannot be the instance of another class. To make disjoint classes such that the object of one class cannot be the object of another class so that consistency will be achieved in mobile web applications to give meaningful information to the farmers. It is done with the "&owl; All Disjoint Classes" in protégé 5.0 beta as shown (Figure 6.2).

The class agriculture system can be shown as a function of the crop, farmer, season, weather, zone, farming equipment, irrigation mechanisms, and so on, which are used to model agriculture ontology. From the code specified above, it can be shown in a mathematical form where class biocide is represented through the symbol bi and similarly other classes are also represented such that,

$$Ag = f(c, fr, di, fe, fer, h, im, p, se, sd, so, sw, st, w) \qquad (6.1)$$

where Ag is agriculture; c, crop; di, distribution; fe, fertilizer; fer, farmer; h, harvest; im, irrigation mechanisms; p, pesticides; se, season; sd, seed; so, soil; sw, sowing; st, storage methods; w, weather.

Furthermore, class farmer is represented through the symbol fer. To illustrate the modeling in agriculture ontology, we take the class crop and establish the relationship with other classes. The relationship is defined as the object property. To develop semantics, we construct class axioms and property axioms.

```
<rdf: Description>
<rdf: type rdf: resource="&owl; AllDisjointClasses"/>
    <owl: members rdf: parseType="Collection">
    <rdf: Description rdf:about="http://www.semanticweb.org/ontologies/2015/2/untitled-ontology-98#Biocides"/>
    <rdf: Description rdf:about="http://www.semanticweb.org/ontologies/2015/2/untitled-ontology-98#Crop"/>
    <rdf: Description rdf:about="http://www.semanticweb.org/ontologies/2015/2/untitled-ontology-98#Farmer"/>
    <rdf:Descriptionrdf:about="http://www.semanticweb.org/ontologies/2015/2/untitled-ontology-98#Distribution"/>
    <rdf: Description rdf:about="http://www.semanticweb.org/ontologies/2015/2/untitled-ontology-
98#FarmingEquipments"/>
    <rdf: Description rdf:about="http://www.semanticweb.org/ontologies/2015/2/untitled-ontology-98#Fertilizers"/>
    <rdf: Description rdf:about="http://www.semanticweb.org/ontologies/2015/2/untitled-ontology-98#Harvest"/>
    <rdf: Description rdf:about="http://www.semanticweb.org/ontologies/2015/2/untitled-ontology-98#Irrigation"/>
    <rdf: Description rdf:about="http://www.semanticweb.org/ontologies/2015/2/untitled-ontology-98#Packing"/>
    <rdf: Description rdf:about="http://www.semanticweb.org/ontologies/2015/2/untitled-ontology-98#Prices"/>
    <rdf: Description rdf:about="http://www.semanticweb.org/ontologies/2015/2/untitled-ontology-98#Season"/>
    <rdf: Description rdf:about="http://www.semanticweb.org/ontologies/2015/2/untitled-ontology-98#Seed"/>
    <rdf: Description rdf:about="http://www.semanticweb.org/ontologies/2015/2/untitled-ontology-98#Soil"/>
    <rdf: Description rdf:about="http://www.semanticweb.org/ontologies/2015/2/untitled-ontology-98#Sowing"/>
    <rdf: Description rdf:about="http://www.semanticweb.org/ontologies/2015/2/untitled-ontology-98#Storage"/>
    <rdf: Description rdf:about="http://www.semanticweb.org/ontologies/2015/2/untitled-ontology-98#Weather"/>
    <rdf: Description rdf:about="http://www.semanticweb.org/ontologies/2015/2/untitled-ontology-98#Zone"/>
    </owl: members>
  </rdf: Description>
```

Figure 6.2 Code for building agriculture ontology.

6.3.3 Building Object Property Between the Classes in OWL-DL

The object property explains two classes, such as crop and season. The classes, crop and season, are associated through the object property "is determined by". It specifies that a particular crop "is determined by" season and, inversely, season determines a particular crop, which is implemented in protégé 5.0 beta and shown in Figure 6.3. The class crop and season are represented through the symbols c and se, respectively. Figure 6.4 shows the usage of the class crop in the Protégé 5.0 beta framework.

The object property can be shown through symbolized form as shown below:

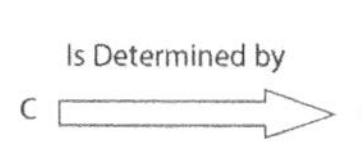

The object property "is determined by" is an asymmetric, irreflexive property and functional property. The domains crop class and range season class are shown in Figure 6.3.

```
<owl:ObjectPropertyrdf:about="http://www.semanticweb.org/ontologies/2015/2/untitled-ontology-98#isDeterminedBy">
<rdf:type rdf:resource="&owl;AsymmetricProperty"/>
  <rdf:type rdf:resource="&owl;FunctionalProperty"/>
   <rdf:type rdf:resource="&owl;IrreflexiveProperty"/>
   <rdfs:domain            rdf:resource="http://www.semanticweb.org/ontologies/2015/2/untitled-ontology-98#Crop"/>
  <rdfs:range             rdf:resource="http://www.semanticweb.org/ontologies/2015/2/untitled-ontology-98#Season"/>
<owl:inverseOf           rdf:resource="http://www.semanticweb.org/ontologies/2015/2/untitled-ontology-98#ditermines"/>
</owl:ObjectProperty>
```

Figure 6.3 Code for building object property.

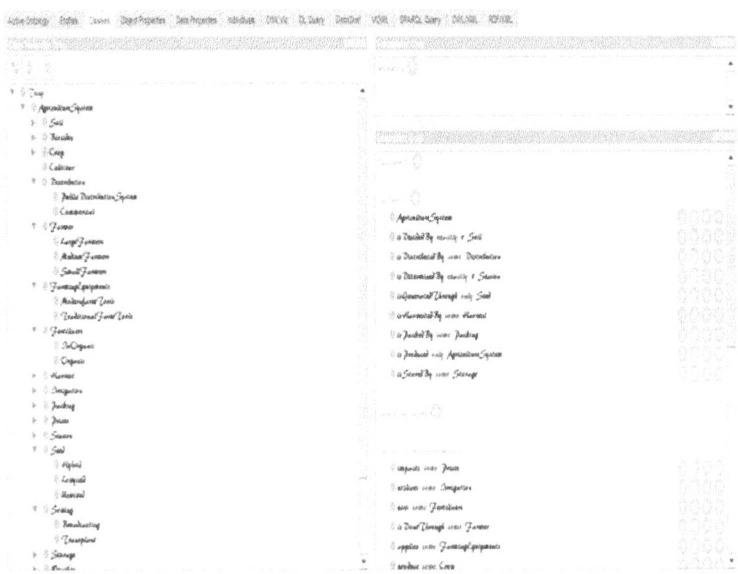

Figure 6.4 Usage of class crop in agriculture ontology.

6.3.3.1 *Building Object Property Restriction in OWL-DL*

The object property "determine" is an asymmetric, irreflexive property and inverse functional property. It has its domains as season class and range crop class, which are shown in Figure 6.5. This object property is a sub-property of the object property "produce," which is held among the agriculture system and crop.

```
<owl:ObjectProperty        rdf:about="http://www.semanticweb.org/ontologies/2015/2/untitled-ontology-
98#determines">
 <rdf:type rdf:resource="&owl;AsymmetricProperty"/>
 <rdf:type rdf:resource="&owl;InverseFunctionalProperty"/>
 <rdf:type rdf:resource="&owl;IrreflexiveProperty"/>
 <rdfs:range rdf:resource="http://www.semanticweb.org/ontologies/2015/2/untitled-ontology-98#Crop"/>
 <rdfs:domain              rdf:resource="http://www.semanticweb.org/ontologies/2015/2/untitled-ontology-
98#Season"/>
 <rdfs:subPropertyOf       rdf:resource="http://www.semanticweb.org/ontologies/2015/2/untitled-ontology-
98#produce"/>
</owl:ObjectProperty>
```

Figure 6.5 Code for defining restrictions in object property.

The class crop is a subclass of the agriculture system. In this agriculture ontology, the property "is decided by" is represented between crop and soil. This meant that the crop is decided by the soil. This agriculture ontology specifies the property restriction as cardinality restriction type for the property or relationship "is decided by" among class crop and season, symbolized as (c) and (se). For example, the crop is decided by one soil type where the soil can be black, red, sandy, alluvial, and laterite type that is described in the agriculture ontology. Hence, it represents the cardinality property restriction as 1 between the crop and season classes in this agriculture ontology, which is shown below. Figure 6.4 shows the class hierarchy of crop class, the object property, and property restrictions between the crop and other classes in the agriculture ontology. To obtain symbolized form, the class crop and season are represented through the symbols c and so, and the relationship "is decided by" among the classes is shown through the arrow. The ontology graph of agriculture ontology is shown in Figure 6.6.

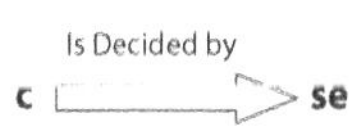

This ontology describes the class crop as a subclass of the agriculture system. It specifies the qualified cardinality property restriction "1" among the crop and season classes. All these are done through the protégé 5.0 beta framework.

6.3.4 Developing Social Ontology

The social elements, such as wikis, podcasting, social network, and content hosting, etc., and these classes are designed through OWL-DL to

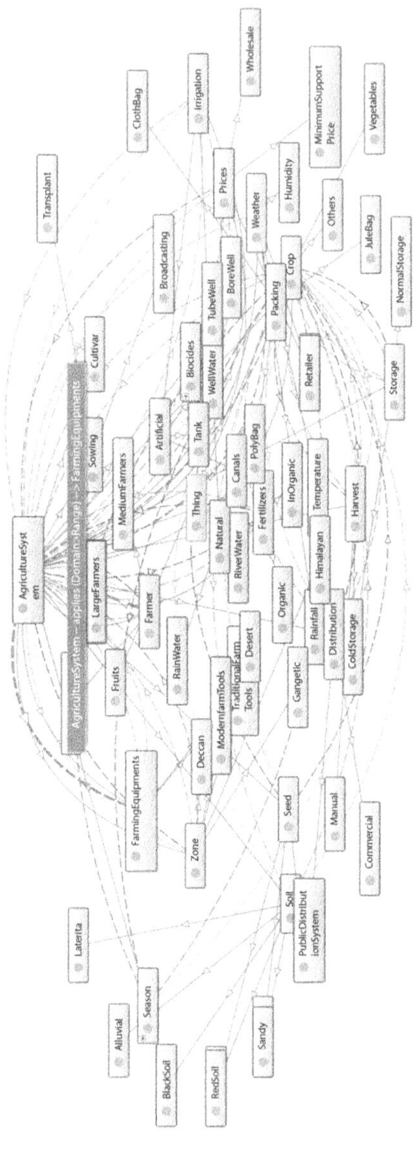

Figure 6.6 Ontology graph for agriculture ontology.

form social ontology. The social elements are the classes, such as wikis, podcasting, social network, and content hosting, and so on, and describe relationships/properties by using OWL. Various class axioms about the classes are established, and class hierarchy is built by using OWL-DL and uses description logic reasoner to check the consistency in the social ontology and compute the class hierarchy. The property restrictions and object property characteristics are also described for a property or relationships that hold among the classes by using OWL-DL in protégé 5.0 beta framework for a social ontology.

6.3.4.1 Building Class Axioms

The social ontology is built by taking the class axiom that Facebook is a subclass of the social network. Also, the class axiom specifies that feed post is a subclass of Blogsphere. Also, we have made all the classes disjoint from each other as shown in Figure 6.7, and the asserted model of social ontology is shown in Figure 6. 8.

Hence, the class social ontology is shown as a function of Blogsphere, Content Hosting, Podcasting, Wikis, Social network, and so on. From the code specified above, it can be shown in a mathematical form where the class wiki is represented through the symbol w, and similarly, other classes are also represented such that,

$$SC = f(B\ , C, P, W, SN) \tag{6.2}$$

where SC is social ontology; B, Blogsphere; P, podcasting; W, wikis; SN, social network.

```
<rdf:Description>
    <rdf:type rdf:resource="&owl;AllDisjointClasses"/>
    <members rdf:parseType="Collection">
      <rdf:Description                    rdf:about="http://www.semanticweb.org/cet-
it/ontologies/2017/11/social-ontology-20#Blogsphere"/>
      <rdf:Description                    rdf:about="http://www.semanticweb.org/cet-
it/ontologies/2017/11/social-ontology-20#ContentHosting"/>
      <rdf:Description                    rdf:about="http://www.semanticweb.org/cet-
it/ontologies/2017/11/social-ontology-20#Podcasting"/>
      <rdf:Description                    rdf:about="http://www.semanticweb.org/cet-
it/ontologies/2017/11/social-ontology-20#SocialOntology"/>
      <rdf:Description                    rdf:about="http://www.semanticweb.org/cet-
it/ontologies/2017/11/social-ontology-20#Socialnetwork"/>
      <rdf:Description                    rdf:about="http://www.semanticweb.org/cet-
it/ontologies/2017/11/social-ontology-20#Wikis"/>
      <rdf:Description rdf:about="&owl;Thing"/>
    </members>
  </rdf:Description>
```

Figure 6.7 Code for social ontology.

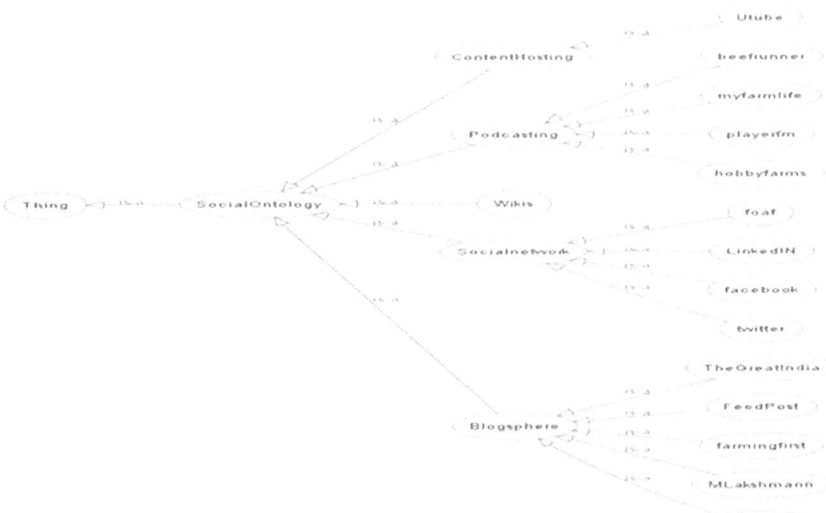

Figure 6.8 Asserted model for social ontology under agriculture domain.

6.3.4.2 Analysis of Social Web Components on Domain Ontology Under Agriculture System

A survey was conducted with multiple-choice questionnaires, and different users (no of participants, 60), such as farmers and vendors, participated in the study. After getting their responses, the data are analyzed and the relation between social elements, web components, users under the agriculture domain is designed. After getting their responses, the result reveals that content hosting and social networking social web applications are more useful for farmers and vendors related to the agriculture domain because they have more good responses than others. Next to blogs and podcasts, social web applications are useful for farmers and vendors (Table 6.1, Figure 6. 9).

6.4 Building Social Ontology Under the Agriculture Domain

6.4.1 Building Disjoint Class

The social web elements, such as wikis, podcasting, content hosting under agriculture domain, specified classes, such as framer, crop, seed, and zone, and so on. At first, making each class have made a disjoint so that an

Table 6.1 Questionnaires and responses of participants.

Questions	Content hosting	Social networking	Podcasting	Wikis	Blogs
Q.1. What are the steps of producing crops?	88%	59%	49%	68%	43%
Q.2. How fertilizers will be placed in corps?	85%	71%	48%	61%	41%
Q.3. What is the best season of producing wheat crops?	89%	67%	52%	65%	47%
Q.4. Which type of soil is best for producing vegetable tomato?	82%	54%	44%	59%	51%
Q.5. What are the storage methods for storing corps and which is best for storing corps?	79%	57%	59%	50%	48%
Q.6. Why do African Youths Ignore the Potentials in Agriculture?	87%	63%	60%	51%	58%
Q.7. What is Commercial Agriculture?	86%	56%	49%	54%	56%
Q.8. What is Smart Agriculture?	80%	57%	46%	55%	41%
Q.9. Is a Graduate Degree in Agriculture Worth it? Why?	77%	64%	44%	57%	44%
Q.10. What is the Difference Between Agronomy and Agriculture?	79%	68%	51%	63%	52%

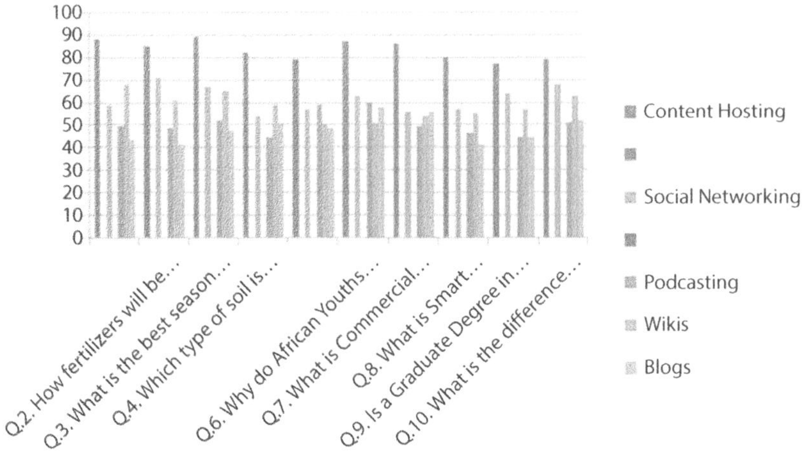

Figure 6.9 Analyzing social elements over agriculture domain ontology.

individual of one class cannot be an individual of another class. After making them disjoint, the combined class hierarchy has been built, which is shown in Figure 6.10.

Here, the class "BlogSphere" disjoint with every class present in the hierarchy to build a social semantic ontology. Here, the social semantic ontology is an integrated ontology that is formed by combing agriculture ontology and social ontology. It can be shown in equation 6.3.

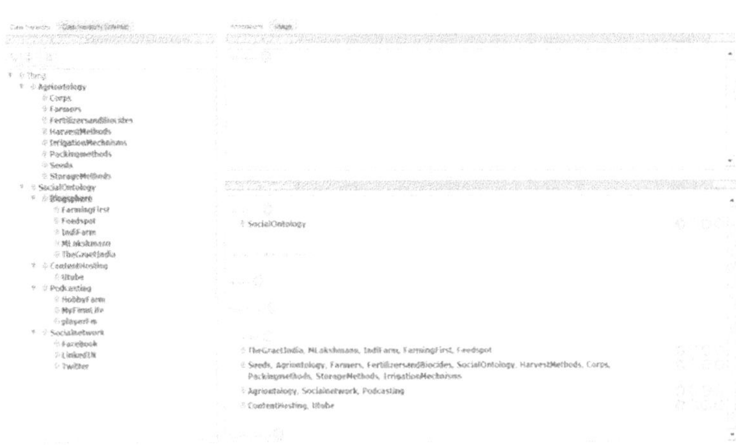

Figure 6.10 Class hierarchy of social semantic ontology under agriculture domain.

$$SMC = f\ (f(Ag)\ \cup\ f(SCO)) = f(f(c, di, fe, fer, h, im, p, pr, se, sd, so, sw, st, w, z)\ \cup\ f(B, C, P, W, SN)) \tag{6.3}$$

where SMC is social semantic ontology, which is a function of these two ontologies as above.

6.4.2 Building Object Property

In analysis, it is shown that producing crops are provided in BolgSphere. The object property has been established, such as "AreProvidedIn," among the subclasses of BlogSphere and Crops. It is defined as Crops and are provided in IndiFarm, MLakshmann, and MyFirmLife. The object property restrictions are specified such that it is an asymmetric, irreflexive, and functional property. This object property has domain class HarvestMethods and range as IndiFarm, MLakshmann, and MyFirmLife classes. Figure 6.11 and Figure 6.12 show the ontology graph of a social semantic ontology under the agriculture domain. To obtain a mathematical form, it represents class crop through symbol c as shown in section 6.3.1 and class Blogsphere through symbol B as defined in section 6.3.2.

The object property "AreProvidedIn" is shown below

```
<ObjectProperty rdf:about="http://www.semanticweb.org/cet
it/ontologies/2017/11/social-ontology-20#AreProvidedIn">
    <rdf:type rdf:resource="&owl;AsymmetricProperty"/>
    <rdf:type rdf:resource="&owl;FunctionalProperty"/>
    <rdf:type rdf:resource="&owl;IrreflexiveProperty"/>
    <rdfs:domain rdf:resource="http://www.semanticweb.org/cet
it/ontologies/2017/11/social-ontology-20#ProducingCorps"/>
    <rdfs:range rdf:resource="http://www.semanticweb.org/cet
it/ontologies/2017/11/social-ontology-20#IndiFarm"/>
    <rdfs:range rdf:resource="http://www.semanticweb.org/cet
it/ontologies/2017/11/social-ontology-20#MLakshmann"/>
    <rdfs:range rdf:resource="http://www.semanticweb.org/cet
it/ontologies/2017/11/social-ontology-20#MyFirmLife"/
</ObjectProperty>
```

Figure 6.11 Code for object property in social ontology.

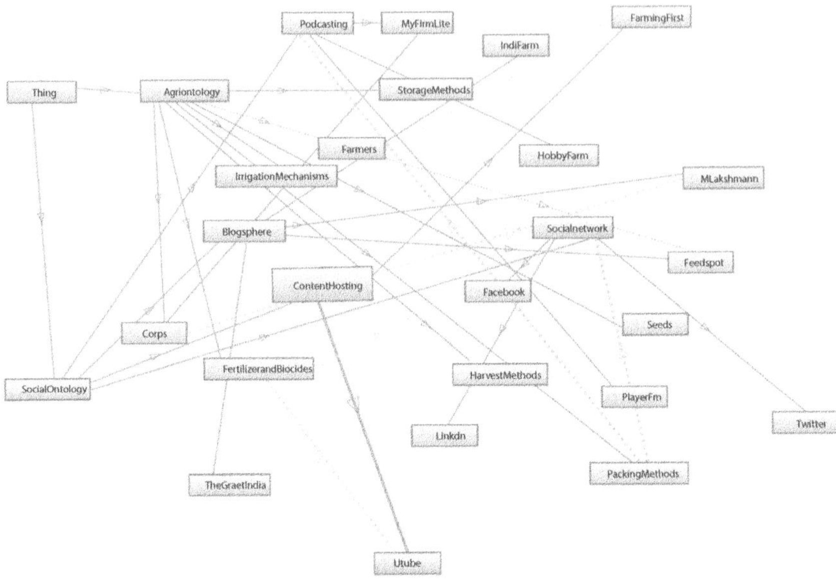

Figure 6.12 Ontology graph for social semantic ontology.

6.5 Validation

The ontology is created by the protégé OWL-DL creator, a user tool called reasoner. The reasoner has used an additional plug-in with protégé OWL-DL that checks and validates for the overall consistency of the created integrated ontology by parsing through each class and class axioms along with property characteristics and restrictions in the class hierarchy, which is present in the agriculture ontology. The ontology is said to be consistent and meaningful if it is successfully classified and validated by the reasoner. The reasoner can be of several versions, which are compatible in protégé 5.0 beta such as pellet, hermi, shet++, and so on. The validation is shown in agriculture ontology in Figure 6.13. Similarly, the social ontology and social semantic ontology are validated in pellet, hermi, shet++, and so on, reasoner in protégé 5.0 beta framework [29].

6.6 Discussion

The semantic ontology has been modeled by taking agriculture systems and social applications. The agriculture system is designed in OWL-DL, and

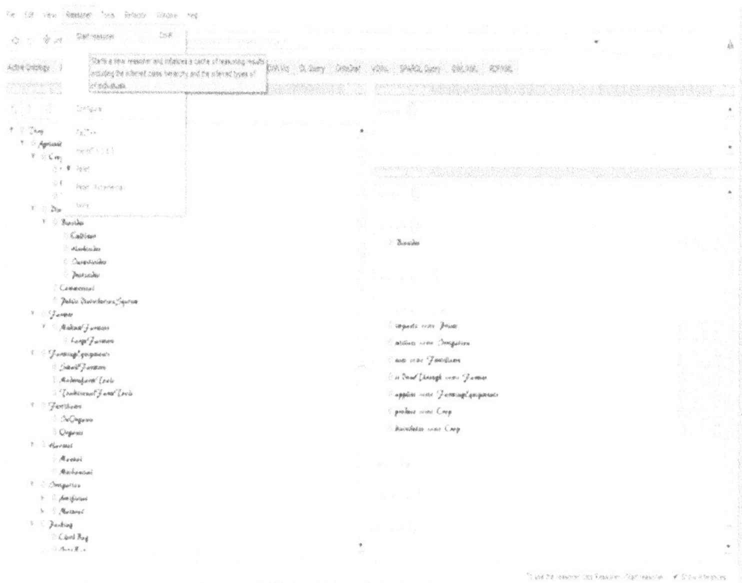

Figure 6.13 Validation of social semantic ontology.

agriculture ontology was developed. The social web application is designed in OWL-DL to build social ontology. We have analyzed the response of questionnaires (multiple choice questionnaires), which are made by farmers and vendors. We have analyzed that content hosting and social network social web applications play a vital role in retrieving information. After that Blogs and Podcasting are having greater importance in finding/retrieving information from the agriculture domain. Hence, these social web components are integrated with agriculture ontology so that when farmer or vendor of an agriculture sector want to search information, then either vendor or farmer can directly be linked or connected with content hosting and social network web applications.

6.7 Conclusion and Future Work

In this chapter, we try to find the effects of social networks over semantic web specified to any domain. The relationships among the social network semantic ontology and semantic web specified domain based on user's perspectives have been analyzed. Social semantic ontology has been built with relationships, which satisfy class axioms and property axioms for analyzing information in a meaningful manner. Again, by using

SPARQL (query language), information can be retrieved from the integrated ontology. An analysis has been done which emphasizes that content hosting and social network social application are more important for retrieving information from the web. After that, blogs and podcasting have greater importance in finding information on domain ontology. These social applications are integrated with agriculture ontology. Hence, these social applications are integrated with agriculture ontology so that when a farmer or vendor of an agriculture sector wants to search for information, he can directly be linked with content hosting and social network applications.

The major contribution of this paper is the usage of the ontology model for the agriculture sector, which is of benefit to particular to farmers for getting the knowledge about crop data by season. From our analysis, it has been found that social web applications, such as social networks and content hosting, are more efficient than other social applications. This analysis is modeled by establishing relationships among domain ontology and social ontology in this study. Hence when farmers want to find information, they can relate to social web components and get information through the social network, content hosting, and blogs automatically.

References

1. Stern, J., *Introduction to web 2.0 technologies*, 2008. [online] www.wlac.edu/online/documents/Web_2.0%20v.02.pdf.
2. Kamani, K. and Kathiriya, D., Cultivate ICT & Networking: The Role of Social Media in Agriculture, in: *Proc. International Conference on Information Systems & Computer Networks*, vol. 37, pp. 1–52, 2013.
3. *Wikis, Blogs & Web 2.0 technology*, 2014. [online] www.unimelb.edu.au/copyright/information/../wikisblogsweb2blue.pdf.
4. Adedoyin-Olowe, M., Gaber, M.M., Stahl, F., *A survey of data mining techniques for social media analysis*, ArXiv preprint arXiv:1312.4617, pp. 1–25, Dec 17, 2013.
5. Bhusan, G.J., TCS m-krishi: TCS Crop Umbrella, Media Lab Asia. *Int. J. Potato Res.*, 44, 2, 121–125, 2010.
6. Basu, D., Das, A., Goswami, R., Accessing Agricultural Information through Mobile Phone: Lessons of IKSL Services in West Bengal. *Indian Res. J. External Educ.*, 12, 3, 102–107, 2012.
7. Ficarelli, P.P. and Glendenning, C.J., *The relevance of content in ICT initiatives in Indian agriculture*, vol. 1180, pp. 1–40, International Food Policy Research Institute Discussion Paper, 2012.

8. Zhdanova, A.V., Ning, L., Moessner, K., Semantic Web in Ubiquitous Mobile Communications, in: *The Semantic Web for Knowledge and Data Management: Technologies and Practices*, pp. 41–62, IGI Global, 2009.
9. Sohaib, A. and Parsons, D., ThinknLearn: An ontology-driven mobile web application for science enquiry based learning. *7th International Conference on Information Technology and Application*, pp. 255–260, 2011.
10. Nan, X., Zhang, W.S., Yang, H.D., Zhang, X.G., Xing, X., CACOnt: a ontology-based model for context modeling and reasoning. *Appl. Mech. Mater.*, 347, 2304–2310, 2013.
11. Ahmed, S. and Parsons, D., COMET: context ontology for mobile education technology. *Artificial Intelligence in Education*, Springer, Berlin Heidelberg, pp. 414–416, 2011.
12. Brunie, L., Ejigu, D., Scuturici, M., An ontology based approach to context modeling and reasoning in pervasive computing, in: *Pervasive Computing and Communications Workshops. PerCom Workshops 07. Fifth Annual IEEE International Conference*, pp. 14–19, 2007.
13. Erétéo, G., Buffa, M., Gandon, F., Corby, O., Analysis of a real online social network using semantic web frameworks, in: *International Semantic Web Conference*, Springer, Berlin Heidelberg, pp. 180–195, 2009.
14. Zhang, C., Jin, Y., Jin, W., Study of Data Mining Algorithm in Social Network Analysis. *3rd International Conference on Mechatronics, Robotics and Automation*, 2015.
15. Jamali, M. and Abolhassani, H., Different aspects of social network analysis, in: *Web Intelligence. IEEE/WIC/ACM International Conference*, IEEE. The Agricultural Promotion and Investment Corp. Of Orissa Ltd. APICOL, Bhubaneswar, Orissa, India, pp. 66–72, 2006.
16. Mishri, B.K., *Country Pasture/Forage Resource Profiles. State of Indian Agriculture. 2013*, Directorate of Economics and Statitics, Ministry of Agriculture, New Delhi, India, 1999, [online] http://www.agricoop.nic.in.
17. Mitra, S. and Sareen, J.S., Adaptive policy case study: agricultural price policy in India, in: *Designing Policies in a World of Uncertanity, Change and Surprise: Adaptive Policy making for Agricultural and Water resources in the Face of Climate Change - Phase I Research Report*.
18. *The Pocket book on Agricultural Statitics*, Directorate of Economics and Statitics, Ministry of Agriculture, New Delhi, India, 2013, [online] http://www.eands.dacnet.nic.in.
19. Alkhammash, E., Mohamed, W.S., Ashour, A.S., Dey, N., Singh, A., Balas, V.E., Designing Ontology for Association between Water Quality and Kidney Diseases for Medical Decision Support System, in: *Conference: VI International Conference Industrial Engineering and Environmental Protection*, 2016.
20. Klotz, B., Datta, S.K., Wilms, D., Troncy, R., Bonnet, C., A Car as a Semantic Web Thing: Motivation and Demonstration, in: *2018 Global Internet of Things Summit (GIoTS)*, IEEE, pp. 1–6, 2018.

21. Suesatsakulchai, A., Buranarach, M., Tetiwat, O., Development of Blood Donor Complication Semantic Retrieval System using the Ontology Application Management Framework. *Int. J. Appl. Eng. Res.*, 13, 5, 2361–2367, 2018.
22. Munir, K. and Anjum, M.S., The use of ontologies for effective knowledge modelling and information retrieval. *Appl. Comput. Inform.*, 14, 2, 116–126, 2018.
23. Abbasi, A.A. and Kulathuramaiyer, N., A systematic mapping study of database resources to ontology via reverse engineering. *Asian J. Inf. Technol.*, 15, 4, 730–737, 2016.
24. Chantrapornchai, C. and Choksuchat, C., Ontology construction and application in practice case study of health tourism in Thailand. *SpringerPlus*, 5, 1, 2106, 2016.
25. Ali, S., Khusro, S., Ullah, I., Khan, A., Khan, I., Smartontosensor: ontology for semantic interpretation of smartphone sensors data for context-aware applications. *J. Sens.*, 2017, 1–26, 2017.
26. Shah, S.K.A., Khusro, S., Ullah, I., Khan, M.A., Semantic Bookmark System for Dynamic Modeling of Users Browsing Preferences, in: *Computer Science On-line Conference*, Springer, Cham, pp. 279–287, 2018.
27. Saravanan, S., Hailu, M., Gouse, G.M., Lavanya, M., Vijaysai, R., Optimized secure scan flip flop to thwart side channel attack in Crypto-chip, in: *Lecture Notes of the Institute for Computer Sciences, Social-Informatics and Telecommunications Engineering, LNICST*, vol. 274, pp. 410–417, 2019.
28. Brahim, N.Y., Mokhtar, S.A., Harb, H.M., *Towards an Ontology-based integrated Framework for Semantic Web*, ArXiv preprint arXiv:1305.7058, vol. 10, pp. 1–9, May 30, 2013.
29. Aghaei, S., Nematbakhsh, M.A., Farsani, H.K., Evolution of the world wide web: From WEB 1.0 TO WEB 4.0. *Int. J. Web Semant. Technol.*, 3, 1, 1–10, 2012.
30. Jamali, M. and Abolhassani, H., Different aspects of social network analysis, in: *Web Intelligence. WI 2006. IEEE/WIC/ACM International Conference*, pp. 66–72, 2006.
31. Bowonder, B., Gupta, V., Singh, A., *Developing a rural market e-hub: the case study of e-Choupal experience of ITC. Planning Commission of India*, 2003.

7

Classification of Normal and Anomalous Activities in a Network by Cascading C4.5 Decision Tree and K-Means Clustering Algorithms

Gouse Baig Mohammad[1]*, S. Shitharth[1] and P. Dileep[2]

[1]Department of Computer Science and Engineering, Vardhaman College of Engineering, Shamshabad, Hyderabad, India
[2]Andhra University, Visakhapatnam, India

Abstract

Cascades of information are a phenomena where individuals take a new action or thought because of their influence. As such technique is transmitted across a social network, broad adoption can occur. In the framework of suggestions and information dissemination on the blogosphere, we are considering cascades of information. Intrusion in a network environment poses a severe security risk. The intrusion detection system in the network is designed to detect attacks or malicious activity in a high-detection network while keeping a low false alarm rate. The system's behavior and flashing systems are monitoring important anomalies in the anomaly detection system (ADS). In this research, we present a method of identification of anomalies with "K-means + C4.5," the method of cascading k-means clustering and the decision tree method C4.5, for classifying anomalous and typical computer network operations. K-Means is the first clustering method for separating training into K clusters with a similarity in Euclidean distance. In each cluster, we create decision structures with algorithms from the decision tree C4.5, indicating a density area of typical or abnormal cases. The Decision Tree illustrates the decision constraints for each cluster by learning the subgroups inside this cluster. We use the findings from the decision tree for each class to get a final conclusion. However, the K-means+C4.5 model is shown to be slightly superior to predict computer network anomalous activities with a rating of 99.2% with true positive rate.

**Corresponding author*: gousebaig@vardhaman.org

Mohammad Gouse Galety, Chiai Al Atroshi, Bunil Kumar Balabantaray and Sachi Nandan Mohanty (eds.) Social Network Analysis: Theory and Applications, (109–132) © 2022 Scrivener Publishing LLC

***Keywords*:** Computer network, anomaly detection, classification, K-means, C4.5 algorithm

7.1 Introduction

Network intrusion detection systems (NIDS), which identify policy violations by network administrators, has become common components in security infrastructures. Intrusion detection systems based on anomalies in the network are now the principal research and development emphasis (A-NIDS). The behavior of the system and the flashing systems monitor major anomalies as an abnormality in the system (ADS). Anomaly detection has recently been used to detect attack on computer networks, malicious computer system activity, and web systems misuse [1].

Current ADS class has been established by machine learning technologies, such as neural artificial networks, fuzzy classification devices, multivariate analysis, and others because of its good accuracy at low error rates. However, the mentioned ADS-related research has disadvantages: the researches develop ways of detection of anomalies using machine learning technologies, such as artificial neural networks, pattern matching, and so on, whereas the current breakthroughs of machine learning have shown a better rendering of selection and cascade of many machines [2].

The topologies in the network, which build many natural and synthetic systems, provide a perfect environment for the development of complex phenomena. A well-studied version of this is found when interactions between system components permit an originally localized effect to spread globally. This is a cascade or avalanche. The failure of technological systems such as e-mail networks, electricity grids, for example, is often caused by a cascade of failure caused by an isolated event. Similarly, the transmission and adoption of innovative or cultural fads of infectious diseases can generate social cascades. The dynamics of cascades has been shown to be sensitive to the model of interaction in the underlying network throughout the last several years. A strong theoretical basis for this dependency is one of the aims of the network theory. For this to happen, it is first of all required to build network models, which are both mathematically correct and which capture the highlights of their counterparts in the world. So far, success in this direction has been limited [3]. The major drawback of most of existing network models in this aspect is their lack of realistic structural patterns, notably lack of significant clustering levels referring to three times the likelihood for connected vertices to form triangles.

Most recent research on data flow and influence through networks has been conducted in the epidemiological and epidemic propagation of the network [3, 4]. Classical disease propagation models are based on a host disease stage: an individual is first vulnerable to illnesses and, if exposed to infection, can become infectious. After the illness has ceased, the individual is recovered or removed. The person is then immune for a set period. There is also room for immunity and a person is again sensitive. Susceptible–infected–regenerated (SIRs), therefore, are disease models in which regeneration is never vulnerable again, and the regenerated host population may be regenerated using SIRS susceptible (renergetic) (SIS) models. Because of a network and several afflicted nodes, the epidemic threshold is examined, i.e., conditions in which the disease is dominated or extinguished. Models for the diffusion of ideas or products that attempt to mimic the process can often be classified into two groups:

- The model threshold [5] is where the threshold t for each node in the network t ∈ [0, 1] is usually taken from a certain distribution of probabilities. At the edge of the network, we also assign wu, V connection weights. A node takes the behavior when it is higher than the threshold of a sum of the weight connection between its neighbors who have adopted the behavior $t \leq \sum_{\text{adopters}(u)} w_u, v$
- Independent cascade [6] wherever a neighbor v of a node u adopts, the node u is likely also $P_{u,v}$. In other words, every time a neighbor of yours buys goods, it is also possible for you to decide to buy it.

7.1.1 Cascade Blogosphere Information

In the blog domain [7] (Figure 7.1), most work was done to remove cascades. Whereas information propagates between blogs, there are relatively uncommon cases of actual cascading behavior. This may be because of the bias in the techniques employed to collect pages and infer relationships in the Web crawling and text analysis. All the suggestions are kept in our database, and we know no records are missing. The product concerned and the time the suggestion was made are linked to each review. Blog space studies either expend a great deal on extracting subjects from posts or take the attributes of blog space as an unlabeled URL chart solely. The blogosphere structure can be captured by numerous different models. Work on the spread of information on the basis of topic [8] has shown that its popularity is stable in time for certain topics ("chatter"), while its popularity is more volatile for the other ones

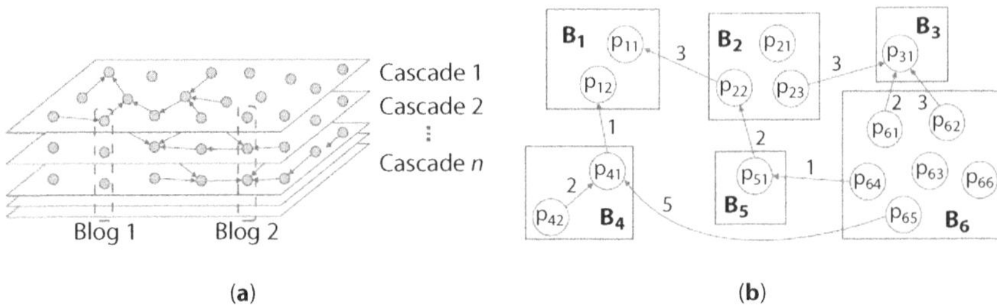

Figure 7.1 Two viewpoints on the development of weblog information cascades. (a) Layer cascade and (b) graph cascade.

("spikes"). Analyze behaviors of the community as inferred from blog-rolls—constant connections between the "friend" blogs. In their extension [9], they analyzed the topological characteristics of connecting graphs in communities and found that a great deal of behavior was characterized by "stars."

7.1.2 Viral Marketing Cascades

The diffusion via the product information network and its acceptance can be described as viral marketing. Research on the impact of social networks on innovation and distribution is carried out in the social sciences in particular. However, this research was generally limited to tiny networks and a single product or service. To learn the network of references, for example [10] interviewed student families taught by the three piano teachers. They discovered that strong ties, those among family and friends, have become more influential than slight ties between acquaintances and are activated for information flow [11]. Word of mouth ads are not limited to peer-by-peer or small-size interactions between people in the context of the internet. Instead, customers can share with everyone their experiences and views about a product. To characterize flows of product information online, quantitative marketing strategies have been proposed [12], and the product and commercial rating has demonstrated the possibility of an item being purchased [13]. More prominent web recommendations allow users to evaluate reviews of other reviewers or rate others directly to set up a trustworthy reviewers' network, which might have very little overlap with an individual's social circle [14, 15] utilized the trusted Epinions reviewer network to develop a viral marketing effectiveness algorithm, assuming that the chance of the individual to purchase a product is dependent on the opinions of the trusted partners in their network.

7.1.3 Cascade Network Building

Social networks can build up many subnetworks with identical nodes connected to various meanings by edges. In addition to the initial network, when information is disseminated among the population, a second layer may be created to reflect the due amount of information [3]. A diffusion/propagation network or a cascade [2] was often referred to. With each node representing a person, we may establish a network, and instead of a Twitter network, each link signals a retweet direction. Thus, if A replays B's tweet, the link between B and A would be made to construct the "retweet network" or "cascade network." In other words, a link would be built. They should be created or inferred to track already existing falls, as mentioned in the last section. Therefore, most early research used various criteria to find cascading networks and establish a connection between two blogs if a link to one is explicit. For instance, no integration mechanisms were observed during the early studies of cascades in blogs.

If no explicit link is provided, it can be inferred with the following functions: the structure of the blog network, history of postings of blogs, text similarities, and timestamps. In most early research, users used their textual credit for the source of information to deduce cascade networks on online social networking networks. CRT, \via," \retweet" and \reshare" are examples of credit allocations [13]. There have also been a lot of attempts to infer cascade networks using social network and timestamps. However, it is feasible to build more accurate networks of cascades with more contextual information. For example [14], employed reshare information, timelines, and feed clicks for inferences and comparisons of networks of cascades with cascades built from monitored information alone.

7.1.4 Cascading Behavior Empirical Research

Whereas the preceding models discuss how processes spread over a network, they are based on presumed influences instead of measured ones. The majority of work has been done in the blog domain on measuring cascading behavior. Blog posts are related by hyperlinks. Because posts are time stamped, the connection patterns to the source may be monitored, and the flow of information to followers can, therefore, be determined from the source [5]. Viral marketing may also be viewed as a network distribution of product knowledge and its adoption [7]. In this case, cascades consist of people who promote things to each other, and therefore, products (and purchases) are disseminated around the network. We observed rich blogosphere and viral marketing behavior [20, 22] and examined several important questions: In actual life, what kind of cascades are common? Are they just like trees, stars, and so on? What are the qualities of the underlying network environment that they reflect? Do some nodes have particular patterns of propagation?

7.1.5 Cascades and Impact Nodes Detection

Cascades can lead to significant insights. Early adopters may, for example, convince their friends to buy this product during viral marketing while trying to sell a product through word-of-mouth effects. The company, therefore, wishes to identify the key nodes to transmit product information over the network [15]. The network outbreaks [21], in which we have a network and a dynamic process that extends through it, detect similar problems and want a number of nodes to identify the process as efficiently as possible. Take the urban water system as an example, which distributes water for families via pipes and interconnections. The contaminants can spread over the network, thus we want to identify several spots where sensors can be installed to efficiently detect pollution. You can describe tasks above, which is a computer issue, as optimization by sets of nodes. However, there has been a decreasing return property termed sub-modularity, and this has been shown. By using sub-modularity, we build [15, 21] almost optimum methods to detect influence nodes and efficiently detect network epidemics.

7.1.6 Topologies of Cascade Networks

A cascade is generally seen as a tree with a single root (the initiator of the cascade) connected with additional nodes. Links to existing nodes in the network cascade can add additional nodes, and all added links have to be kept in a precise temporal order [4]. Nevertheless, cascades are not necessarily tree-like. Their structure varies according to the content kind. Classified cascade networks are information-shared networks used to communicate information between customers and registries with revolutionary technology. The topology of the created cascade network is not specified in this classification. Particularly, with regard to their topology, we consequently offer a different category of cascade systems. The content type and dispersion technology given by the platform are based on this categorization.

The construction of cascading networks that was employed in study involves two major methodologies. The topologies obtained from each strategy are illustrated in Figure 7.2. First, collective cascades, in which a huge cascade network is established, collectively connect individuals to a collection of

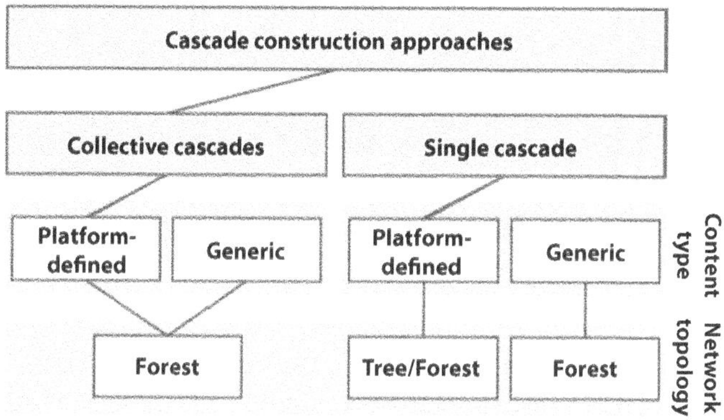

Figure 7.2 The development and topology of cascade networks.

cascade items based on their sharing behaviors (retweet/reblogs). A forest that has several elements is the topology of this network. These huge networks are helpful in examining the patterns of sharing activity inside a platform [14]. The collective cascade networks are usually assessed according to how many times the association between two nodes is established [1].

The second technique is for individual cascades where cascades are separately shared for each item. The first is a platform-led aspect comprising two kinds of material, for example a Twitter tweet or a Tumblr post. The second group (generic elements) covers all the element, such as a URL, a hashtag, a paragraph, or a photo, that can be placed into a platform. Different forms of content demand distinct techniques of data gathering and processing, and an entirely different network structure is created.

A post on Tumblr and Facebook, for example, or a tweet on Twitter, can be shared by plat specific portions. This type of content is distributed deliberately through tools such as retweeting, sharing, and reblogging. Its spreads produce waterfalls which the platform can track or infer. Cascades are based on the flux of user data which may or could not be related to one other through a social chart [5, 6, 18, 19]. These cascade networks follow a topology of the tree perfectly. The source (author) is rooted in the social network and information passes from it. However, certain data may often be omitted since it has been removed owing to limited access to the platform and due to a lack of data, the architecture of the created cascade network is a forest where individual components for each isolated section not connected with the main tree are provided [6].

Because there are no explicit diffusion functions in social networks to disseminate generic elements, such as hashtags or URLs. Time stamps are so typically used to indicate the diffusion of users if these people have a social

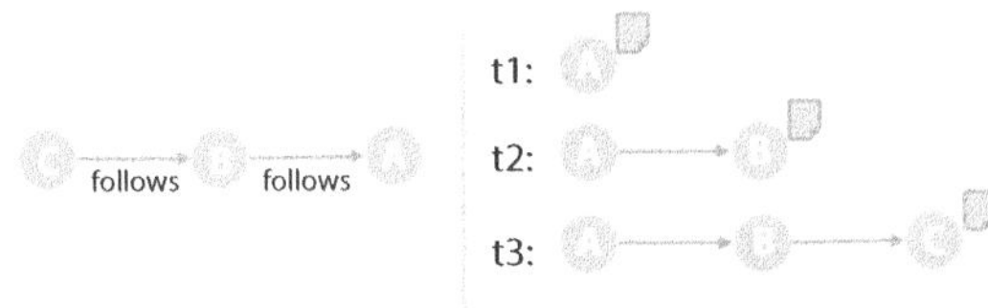

Figure 7.3 Links to cascades, left: perspective on relationships, right: perspective on information flow.

relationship in the social network graph. Cascade generic objects' networks differ from one story's cascade networks (Figure 7.3). These networks involve several introductions into the network, and so their topology is naturally a forest with different components (subcascades). Thus, as structural elements of these networks, the number and size of the subcascades [24] can be utilized.

Collective cascades can simply be turned into a single cascade network through separation of the several branches of the network where the same tale is connected (message). For example, cascade networks are formed according to two blogging methodologies. If you credit one another, then you build a postal network that links posts. They established a blog network from the postal network by compressing and assigning weight to the link between blogs. They built separate cascading trees from the post network according to this approach.

The proposed work contains a supervised approach called as "K Means + C4.5" for identifying abnormalities, constructed in waterfall using two machine learning algorithms (1) the clustering of k-means and (2) the decision tree C4.5. K-means are grouped on workouts to obtain k disjoint clusters in the initial phase. Each K Mean cluster displays an area of comparable examples of the Euclidean distances to its clustering centers. We selected the K Means cluster, for (1) the data approaches are driven by relatively small distribution, and (2) the greedy search strategy ensures that the criteria work at least locally and therefore speeds up the convergence of groups to large quantities of data. In the second phase of K-means+c4.5, the C4.5 with decision-making trees based on the occurrence of each K-mean class would cascade k-means technological technology. The initialized K-means approach underestimated natural groupings under the training data with a small K value. The overlapping groups are, therefore, not recorded in one cluster, and each group is compacted to one. These "forced assignments" can increase the incorrect positive rate or decrease the accuracy of the detection in anomaly detection. Class dominance is the second difficulty in a cluster with few occurrences of a certain class and few other classes when the training data are present. These clusters are weakly associated with other classes dominated by a single class. The aforementioned statistics demonstrate that it is better than separate implementations to cascade the algorithms for machine learning. Two components of the choice process are k-medium and C4.5: (1) selection and (2) categorization. During the selection phase, the cluster closest to the test instance is picked. For this cluster, the decision tree is designed for the cluster supplied. The test instance that employs the findings of the decision trees is classified as normal or abnormal and is labeled as a standard or anomalous cluster on the classed label.

We conduct experiments using nonlinear component analysis methods based on the network anomaly data taken from the MITDARPA 1999 network traffic. The data collection contains aberrant and normal computer network domain behavior patterns. Six measures are used to assess performance of the K-Means+C4.5 cascading technique:

1. precision or true-positive rate of detection (TPR),
2. false-positive rate (FPR),
3. accuracy,
4. total accuracy (or accuracy),
5. F-measure, and
6. curves and regions under ROC curves for receiving operating function.

7.1.7 Proposed Scheme Contributions

The contributions to the proposed scheme are set out as follows: this research offers a new technique for categorizing k-means and decision-making forum to address the issues of compulsory assignment and class supremacy in k-means for normal and healthy computer system data classification. A fresh technique is presented in the essay. The research evaluates and analyzes K-Means+C4.5 and C4.5, utilizing the six performance metrics, for approaches to the decision tree. The research presents a new technique of combining two effective approaches for improvements in categorization. The paper proposes a high-performance anomaly detection system from an anomaly detection viewpoint.

The rest of the paper is arranged as follows: in section 7.2, a brief related survey is provided. In section 7.3, we describe the methods of anomaly detection in the k-means and C4.5 decision tree. In section 7.4, the anomaly detection method K-Means+C4.5 is presented. The experimental data sets are discussed in section 7.5. We conclude our work in section 7.6 and provide guidance for the future.

7.2 Literature Survey

A comprehensive analysis of anomaly detection systems and applications based on categorization will be provided. Anomaly detection methods are based on classification. A classification is used to learn from a number of models labeled and then categorize your test case into one class with a model (testing) [5] The approaches for classification may be separated into two phases: (1) a period of training and (2) a phase of testing. During the training phase, the classifier is trained using the label details. In the test step, test instances are classified by a classification algorithm as normal or abnormal (anomaly). Anomaly methods are classified into two primary classes, (1) anomaly classification techniques and (2) classification-based detection methods. Anomaly classification methods are examples of many ordinary classes are the training data in the first technique [26]. Any regular class in the remainder of the class is capable of detecting such abnormality. An anomalous test instance is considered if one of this classifiers does not categorize them as normal. In training, just one class designation is applied to the second method. To learn a discriminatory boundary in the typical cyclic, this approach utilizes a one-class classification algorithm, like a support vector machine (SVM) of one class [5] or a one-class Kernel Fishe Discriminate [24]. Any test instances that are not within the limitations learnt are labeled as abnormal. Anomalies were detected in the Bayesian multiclass network. The potential of class labeling for the individual test data instance is a commonly recognized essential way of a universal category data set employing a naïve Bayesia Network. In this situation, the class label is picked as the projected class with the highest margin. The estimates for each class and the preceding class probability are the probabilities for the observation of the test examples. The core method allows the generalization of categorized data by adding post-test probability per characteristic for each test instances and by allocating a label with the aggregated value to the test instance.

Several variations of the fundamental technique have been created for the network intruders' detection [2, 4], video monitoring news detection [4], text data anomaly, and disease outbreak detection. Various strategies for capturing dependence among qualities using more complicated Bayesian networks have been suggested [3, 10]. Support for the anomaly detection in a class setting has been used with the SVM. Such algorithms employ SVM [22] one-class learning techniques and learn an area which includes data training instances. The SVM is being used in Branasten *et al.* for supervised intrusion detection, [19]. To understand complex areas, kernels, such as the radial basis function (RBF), might be used. The basic technique selects the experimental instance in the learning region for each test instance. If the study area is determined to be usual, it is otherwise labeled aberrant.

The robust SVM (RSVM) is used by Callaway *et al.* [25], and it is robust to the existence of training data defects. For detection of intrusion in system call [9], RSVM was implemented. Anomaly detection technique is based on the learning rules in capturing a system's regular behavior. There are two stages in a basic, multiclass rule-based method. First step is the rule algorithms for learning from training data (such as RIPPER, decision-making bodies, etc.). The second stage is the rule best for each test instance. The second step is the rule. The converse is the oddity of the test instance with the best rule. The fundamental-regulatory technique has been proposed in several smaller forms [6, 8, 11]. For one class anomaly detection, association rule mining [37] was utilized by constructing data rules in an unmonitored manner. Association rules from a categorical set of data are constructed. A support criterion is used to trim the rules with low support and to guarantee that the rules are consistent with strong patterns [27]. First step is the rule algorithms for learning from training data (such as RIPPER, decision-making bodies, etc.). The second stage is the rule best for each test instance. The second step is the rule. The converse is the oddity of the test instance with the best rule. In the middle step of association rule mining algorithms, frequent item sets are formed. Duda *et al.* [30] present a categorical data set anomaly identification algorithm where the anomaly of a test instance is equal to the number of common items in a test instance.

Helmer *et al.* [33] have defined network resilience as adaptation to internal or external defects, which can change the structure of the network so that normal services can be provided. The common component of these definitions implies that the network provides appropriate network services when effectively connected. Most study in network sciences investigated the resilience of the network to reflect the degree of failure tolerance in a network, which is quantified by the huge component. This idea has been used to identify network disruption [2, 8, 22]. The percolation theory is described below. On the other hand, computer scientists examine network resiliency more broadly by taking network secrecy into consideration [13, 14, 33]. The topic of network resilience as regard fault tolerance, adaptability, and recovery was particularly highlighted by a complete literary study in numerous different disciplines [14, 33]. The resilience of a system that functions well can be described by: (1) delivery of the usual service with failure or assault(s), (2) adaptation of system configurations to sudden changes (e.g., faults or annexes) for the maintenance of an ordinary system state (e.g., adaptability). Other research also examined the tolerance and reliability resilience of the network [11, 13, 23, 28, 35], fault tolerance [2, 83], and regenerability.

The results of other studies were also discussed. Fan *et al.* [30] suggested modeling to improve transport network resilience and efficiency inspired by the theory of percolation in the areas of urban road systems that take resilience into consideration. He *et al.* [31] provided a generic analytic theory characterizing the effect on the network structure and sizes of a huge component of color-dependent bond percolation. In the context of the reconnecting likelihood reflecting recoverability, Muniyandi *et al.* [36] presented an approach to prevent or lessen the full collapse of a system of interdependent networks encountering cascade failures.

Janikiram *et al.* [35] developed a probabilistic site percolation solution based on the creation of community-based functions for a wireless sensor network. Yuan *et al.* [54] have studied k-core percolation when the node is failing in random, localized, or targeted attacks on random and scale-free network models while losing connections on the basis the threshold k is given. Network adaptability refers to the capacity of a network to shift the topology of the network, such as edges adaptation or redundancy to handle rapid system and environmental changes seamlessly [28]. The network adaptability is investigated in a way which has a very tight connection between the efficacy of the reaction to abnormal conditions and the reliability and effectiveness of malicious behavior [28, 35]. To handle advanced methods to attacks, the network must also be capable of deploying adequate defense mechanisms [10, 11]. The availability of networks was examined to show the level of disaster management network resilience [23, 43]. As regard the abovementioned publications, our work adopts a network science approach to the analysis of network behaviors under attack by using percolation theory. However, in contrast to network science research, which generally considers network resilience to be a failure tolerance, we extend to a more fault tolerance the concept of network resilience.

The adaptability of the network to cascade due to targeted attacks has not been examined to this day. The theory of percolation network resilience is typically examined by measuring the gigantic component using percolation theory. It is often used for assessing network connectivity as a major indicator to reflect the degree of network resilience. In this line of network resilience research based on percolation theory, the implications of various sorts of attacks, such as random attacks or target attacks [34], are investigated [25]. The location and the connection are linked to node and rim removals. By picking the initial number of nodules or edges from the network following an assault, the percolation impact on the dimensions of the largest and most dominant is measured [36, 37]. In most network science methodologies, a critical percolation threshold as a measure of network resilience was identified [4, 38]. Different selection algorithms for node or borders have been devised to mimic targeted attacks, such as degree or wear and tear, depending on a node centrality measure.

Contrary to the network science method, the concept of network resilience to managing network services was explored by computer scientists. The percolation theory has been used for the study of a critical occupation chance (i.e., how many nodes exist within networks) to identify epidemics leading to a cascade failure [12, 22], the dimensions and cost-effective methods of immunization within an enterprise network [28] and the percolative theory have been used to identify critical occupation likelihood. Baribasi *et al.* [17] used percolation theory to achieve a lower bitrate bound in wireless network sensor settings per source-destination pair. Blume *et al.* [9] have identified an ideal framework for strong multifaceted routing of networks or node disconnections. However, neither of the abovementioned projects addressed cyber-attack failures. Although several of the aforementioned works address network resilience by taking into account different types of attack behavior, network resilience is mostly investigated on the basis of fault tolerance.

Keerthi *et al.* [38] introduced a method of degradation that turns the issue of quadratic programming into a sequence of quadratic issues, improving study rates and reducing the requirements for memory space without decreasing classification precision. The proposal includes an SMO algorithm, which can resolve each suboptimal problem's solution and speed up algorithm convergence. However, the computing complexity of the process is increased. Collobert *et al.* [39] further increased the algorithm's learning speed. To make the system more efficient to detect, a parallel learning method has been suggested for sample segmentation that splits the sample into P subsamples using the concept of divide and conquer. However, if the P sets are divided randomly, the accuracy of the classifications decreases and extra rules increase its complexity. However, classification accuracy is diminished, and other constraints add to the complexity of the procedure, which minimizes computational complexity when P subsets are divided randomly.

Cataltepe *et al.* [40] propose a semisupervised decision-tab technique online feature selection that leverages online clusters to summarize the network data available using the extensive clusters functions for the clusters. Each cluster will be marked as abnormal or normal, based on the decision tree input characteristics and the originals characteristics, as well as the relationship between the class description, depending on the characteristics of the decision tree, depending on the decision tree. Although the algorithm's computer complexity has decreased, the detection rate is not yet high. A better anomaly detection technique is suggested in this study that combines kmeans with C4.5. This algorithm's main idea is to first group each category by K-means, then create numerous C4.5 hyperspheres based on the results of the clustering. By computing the sample data, the sample's affiliation to the minimal superficial ranges created by C4.5 is established. Lastly, the KDD CUP99 data set is used to simulate the technique proposed for detection.

7.2.1 Network Failures

The following characteristics may create network failings [33]: network failures based on connecting, cascading, and functionality. Although these three categories are categorized by us, they are interwoven because nodes can be overwhelmed, owing to disconnections from other nodes or functional failures of other nodes, which could lead to a cascade failure too. A network malfunction occurs when a particular portion of connected, effective nodes cannot be maintained by the network. The percolation theory uses this kind of failure to characterize the collapse of the gigantic component after elimination of a significant portion of the nodes or edges. The removal of nodes is selected with the model of an attack, such as random assaults or attacks. Common network architectures, such as the Cataltepez Random Network [16], the Helmers Free Network [33], and the Watts-Strogatz Small Network [15], have been implemented to support the system. Failure of other nodes causes one or more nodes to fail, which could lead to failure of others [42]. A cascade-based failure of the network occurs. The epidemic of transmission of infectious disease is frequently seen as a process of transmission. In the investigation of financial institutions, grids, and communications networks, epidemic models were used to analyze failures besides diseases. It was also used to describe the propagation and compromise of malware or node capture in the cybersecurity domain [6]. Many factors, including behaviors of node neighbors, the area of failure effects, and the chance of failure spread (i.e., infection rate) [33], have an impact on network cascade failures. This malfunction resulted in the failure to offer normal network services. Chau *et al.* [20] evaluated the effect on the criticality and/or interdependence of network nodes of functionality-based failure. The criticality of a node is often quantified by many types of core metrics [38]. He *et al.* [31] further pointed out the primary effect of node functionality failures to be cascade failures. Failures overloaded because of malfunctions of the node [32] can also lead to network malfunctions, depending on the functionality. A failure of a node can lead to cascade failures because it can influence the service supply of its connected subcomponents overall [3]. To undertake a complete study of our suggested network adaptation techniques, we take into consideration all sorts of the abovementioned failures of the network because of distinct node failures (function, overload and security failures), noninfective (non-infectional), and assault methods (random vs. targeted).

7.3 Methodology

7.3.1 K-Means Clustering for Anomaly Detection

The K-means clustering algorithm is a traditional partitioning approach. It splits n items into K clusters with K as input parameters, enabling independent and comparable clusters, and averaging cluster similitude [10]. Although the K-means technique is compact and computer-complex, the algae is vulnerable to outliers [11]. Although it is a computational complexity, because a big extreme value of the subject can considerably distort the data distribution, the square error function is used to determine the convergence together with the K-means method, making it susceptible to the outermost rise, resulting in clustering inaccurate findings.

To calculate the distance (i.e. similarity) between two objects, it requires a distance function. The Euclidean function is the most often used distance function:

$$d(x,y)=\sqrt{\sum_{i=1}^{m}(x_i-y_i)^2} \tag{7.1}$$

where x = $(x_1, \ldots, x_m)$ and y = $(y_1, \ldots, y_m)$ are two m quantity input vectors. All features contribute to the function value equally in the Euclidean distance function. As distinct characteristics are normally measured by different metrics or at different scales, however, before the distance function applies, they need be normalized. The Mahalanobis distance function is an alternative to the Euclidean distance, using the reverse covariance matrix S-1 to express statistical correlations between various characteristics.

$$d(x,y)=\sqrt{(x-y)^2 S^{-1}(x-y)} \tag{7.2}$$

However, for feature vectors with a wide range of dimensions, calculation and reverse covariance matrix are computationally required.

The k means data group N, which shows that k is an advanced parameter, point to k different clusters. The steps of the k-means cluster-based anomaly detection approach follow.

Step 1: Select k random instances from the training data subset as the centroids of the clusters C1; C2; …Ck.
Step 2: For each training instance X:
  a. Compute the Euclidean distance $D(C_i,X),i = 1\ldots k$
  b. Find cluster C_q that is closest to X.
  c. Assign X to C_q. Update the centroid of C_q. (The centroid of a cluster is the arithmetic mean of the instances in the cluster.)
Step 3: Repeat Step 2 until the centroids of clusters C1; C2; …Ck stabilize in terms of mean-squared-error criterion.
Step 4: For each test instance Z:
  a. Compute the Euclidean distance $D(C_i,Z),i = 1\ldots k$. Find cluster C_r that is closest to Z.
  b. Classify Z as an anomaly or a normal instance using the Decision tree.

7.3.2 C4.5 Decision Trees Anomaly Detection

C4.5 is the decision tree for the characteristics of both category and constant. C4.5 splits the attribute values into two partitions based on the chosen threshold, so that all values over the threshold are managed as a single value and a remaining value. The attribute values are also handled missing. C4.5 employs the entropy and information gain for a decision tree as an attribute selection metric. It eliminates the bias of the acquisition of knowledge if an attribute has many result values. The measure of disturbance or impurity is entropy.

$$Entropy = -\sum_{i} P_i \, Log_2$$

Pi is class likelihood. The data gain shows us how crucial it is to have a specific vector attribute. This acquisition of information is used for decision making. A sample decision tree for use in computer networks can be seen in Figure 7.4.

With a number of examples in S, C4.5 creates an initial tree with the algorithm divide and conquer:

- If S are all small or of the same class, the tree is a leaf with the most common class in S.
- Otherwise, select a test with two or more results based on a single property. Set this test to one tree root with a branch

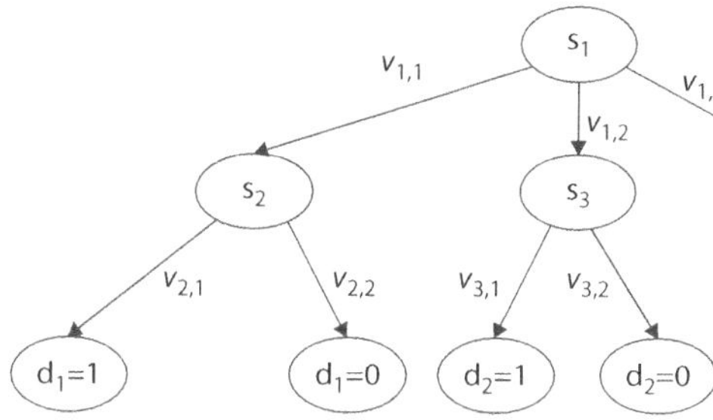

Figure 7.4 A sample decision tree for use in computer network.

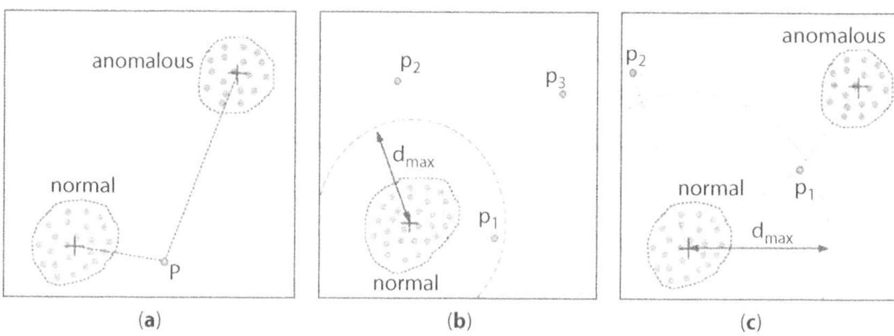

Figure 7.5 (a) Outlier detection classification K = 2, (b) and (c) combined classification and outlier detection classification.

for each test result, divide S into the appropriate subsets, S1, S2, ... in accordance with the result of each case and apply the same technique to each subset recursively.

There are often several tests to select from in this final phase. In two heuristic tests, the information gain that minimizes the total entropy of subsets {Si} (although heavily skewed in the direction of comprehensive testing) and the default dividing ratio of information obtained by the findings are utilized (Figure 7.5).

7.4 Implementation

Anomaly Detection Method Using K-Means+C4.5

7.4.1 Training Phase Z_i

A training data set (x_i, y_i), $i = 1, 2, 3, \ldots, N$. I = 1, 2, 3, … , N in which X I is a continuously valued n-dimensioned vector, and Yi {0,1} represents the respective label of the class "0" as normal and Yi (1) as an abnormality.

There are two steps in the suggested method: (1) training and (2) testing. During training, the initial part of the training space into K disjoints C1, C2, C3, …, CK is done by steps 1 to 3 of k-means–based anomaly detection process. Each k-means cluster was formed via the decision tree C4.5. The k-means approach assures that only one cluster is connected to every training instance. However, in the event of any subgroup or overlap in a group, the decision tree C4.5 trained on this cluster will sharpen the limitations of decisions by dividing instances into space rules.

7.4.2 Testing Phase

We have two distinct phases in the testing phase (1) phase of selection and (2) phase of classification. In the selection phase, calculate the Euclidean distance and find the nearest cluster for each testing case, calculate the cluster decision tree, apply the Zi test instance on the C4.5 decision tree of the closest cluster in the classification phase and classify the test instance abnormality. Below is the algorithm for the approach.

K–Means+C4.5 Algorithm

Selection Phase
Input: Test instances Z_i, $i = 1,2,3, \ldots..,N$.
Output: Closest cluster to the test instance Z_i.
***Procedure* Selection**
Begin
Step 1: For each test instance Z_i
  a. **Compute the Education D(Z_i, r_i),j=1…k, and find the cluster closest to Z_i**
  b. Compute the C4.5 Decision tree for the closest cluster.
End /*End Procedure*/
Classification Phase
Input: Test instance Z_i
Output: Classified test instance Z_i as normal or anomaly
***Procedure* Classification**
Begin
Step 1: Apply the test instance Z_i over the C4.5 decision tree of the computed closest cluster.
Step 2: Classify the test instance Z_i as normal or anomaly and include it in the cluster.
Step 3: Update the center of the cluster.
End /*End Procedure*/

7.5 Results and Discussion

In this section, we offer a performance study with the associated classification algorithms for the suggested approach. To perform relevant experiments, we utilize the well-known data from the 1999 KDD Cup (KDD99) [30]. First, the proposed C-means and the C4.5, cascade algorithms, which include K-means, [4]. There have already been experiences with the ID3 decision tree [16, 18], the algorithms of Naïve Bayes and K-NNN (K-neighbors transductive confidence machines). [1] We are testing the appropriate classification algorithms widely employed in the anomaly detection system of the network. Second, the analysis is validated by comparison of the proposed algorithm.

7.5.1 Data Sets

This section discusses the experimental data set for the detection of abnormalities (KDD99 data set). The Data Set [30] KDD99 is separated into four groups which include DoS (Denial of Service), U2R (Root User) and R2L. (Remote to Local). Connects were summarized by packet data from the original TCP dump files. The KDD99 data set comprises 41 functions for each instance. The KDD99 data collection has taken into account 15000 training samples. The data set for training will consist around 60% of normal data and approximately 40% (anomaly data). For the test, 2500 test instances from the KDD99 data set were selected randomly.

7.5.2 Experiment Evaluation

The testing step was carried out using the KDD99 data set [30] (41 functions), because no functional selection processes have been applied (data set will contain 41 features). We use Weka (Weka 3.5) [29] for an open-source learning framework. Weka is a collection of algorithms for machine learning applications in data mining. We used this tool to compare our method with the other categorization algorithms associated with it [41, 44].

7.6 Conclusion

Most systems for detecting anomalies are designed according to data instance availability. Many techniques for anomaly detection has been developed, whereas others are more generic. This work contains a cascaded algorithm with K-mean and C4.5 algorithms for supervised

Table 7.1 Performance evaluation for the KDD99 data set without using the Feature Selection algorithm [34].

Machine learning algorithms	Performance evaluation in %				
	True positive rate	False positive rate	Precision	Accuracy	F-measure score
K-Means	98.2	2.8	91.4	90.5	83.1
ID3	96.1	4.9	92.2	92.1	92.4
Naïve Bayes	96.0	4.0	94.6	92.1	92.4
K-NN	97.9	3.1	92.2	92.2	92.6
SVM	97.5	2.5	91.6	94.5	91.5
TCM-KNN	98.8	2.2	95.8	96.4	94.6
K-Means + C4.5	99.2	0.8	97.3	94.2	95.2

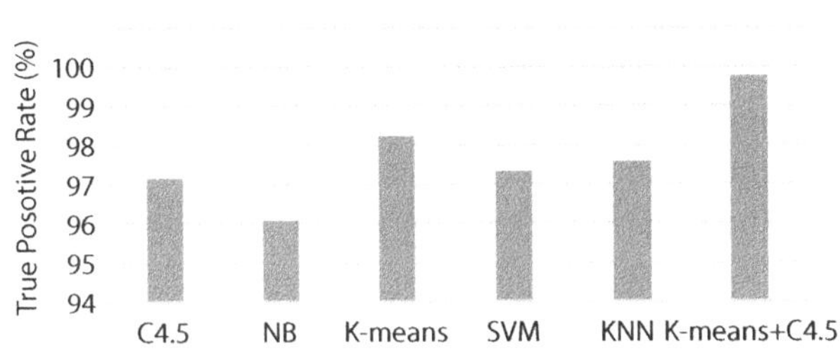

Figure 7.6 Performance measurement index in graphical representation.

abnormality detection. Anomalies in the monitored data set are identified by the proposed algorithm. We use KDD99 data collection for experiments. Performance tests are measured in five measurements: (1) TRR, (2) false FPR, (3) precision, (4) total precision (TA), (5) F-Methods. Measuring is measured by five measures (FM). The suggested approach delivers outstanding detection accuracy to the experimental data (Figure 7.6).

References

1. Mohammad, G.B. and Shitharth, S., Wireless sensor network and IoT based systems for healthcare application. *Mater. Today: Proc.*, 1–8, 2021.

2. Mohammad, G.B. and Kandukuri, P., Detection of Position Falsification Attack in VANETs using ACO. *J. Int. J. Control Autom.*, 12, 6, 715–724, 2019.
3. Shitharth, E. and Prince Winston, D., An Enhanced Optimization algorithm for Intrusion Detection in SCADA Network. *J. Comput. Secur.*, 70, 16–26, 2017.
4. Shitharth, D. and Winston, P., A New Probabilistic Relevancy Classification (PRC) based Intrusion Detection System (IDS) for SCADA network. *J. Electr. Eng.*, 16, 3, 278–288, 2016.
5. Sangeetha, K., Venkatesan, S., Shitharth, S., Security Appraisal conducted on real time SCADA data set using cyber analytic tools. *Solid State Technol.*, 63, 1, 1479–1491, 2020.
6. Thirumaleshwari Devi, B. and Shitharth, S., An Appraisal over Intrusion Detection systems in cloud computing security attacks, in: *2nd International Conference on Innovative Mechanisms for Industry Applications*, p. 122, 2020.
7. Sangeetha, K., Venkatesan, S., Shitharth, S., A Novel method to detect adversaries using MSOM algorithm's longitudinal conjecture model in SCADA network. *Solid State Technol.*, 63, 2, 6594–6603, 2020.
8. Mohammada, G.B., Shitharth, S., Kumar, P.R., Integrated Machine Learning Model for an URL Phishing Detection. *Int. J. Grid Distrib. Comput.*, 14, 1, 513–529, 2021.
9. Glance, N., Hurst, M., Nigam, K., Siegler, M., Stockton, R., Tomokiyo, T., Deriving marketing intelligence from online discussion, in: *11th ACM SIGKDD International Conference on Knowledge Discovery in Data Mining*, pp. 419–428, 2005.
10. Goldenberg, J., Libai, B., Muller, E., Talk of the network: A complex systems look at the underlying process of word-of-mouth. *Mark. Lett.*, 3, 12, 211–223, 2001.
11. Granovetter, M., Threshold models of collective behavior. *Am. J. Sociol.*, 83, 6, 1420–1443, 1978.
12. Granovetter, M.S., The strength of weak ties. *Am. J. Sociol.*, 78, 1360–1380, 1973.
13. Kumar, P.R., Wireless Mobile Charger using Inductive Coupling. *J. Emerg. Technol. Innov. Res.*, 5, 10, 40–44, 2018.
14. Hethcote, H.W., The mathematics of infectious diseases. *SIAM Rev.*, 42, 4, 599–653, 2000.
15. Kempe, D., Kleinberg, J.M., Tardos, E., Maximizing the spread of influence through a social network, in: *9th ACM SIGKDD International Conference on Knowledge Discovery and Data Mining*, pp. 137–146, 2003.
16. Kumar, P.R. and Ananthan, T., Machine Vision using LabVIEW for Label Inspection. *J. Innov. Comput. Sci. Eng.*, 9, 1, 58–62, 2019.
17. Barabási, A.-L. and Albert, R., Emergence of scaling in random networks. *Science*, 286, 5439, 509–512, 1999.
18. Blume, L., Easley, D., Kleinberg, J., Kleinberg, R., Tardos, É., Which networks are least susceptible to cascading failures, in: *IEEE 52nd Annu. Symp. Found. Comput. Sci.*, pp. 393–402, 2011.

19. Brannsten, M.R., Johnsen, F.T., Bloebaum, T.H., Lund, K., Toward federated mission networking in the tactical domain. *IEEE Commun. Mag.*, 53, 10, 52–58, 2015.
20. Callaway, D.S., Newman, M.E., Strogatz, S.H., Watts, D.J., Network robustness and fragility: Percolation on random graphs. *Phys. Rev. Lett.*, 85, 25, 5468–5471, 2000.
21. Chau, C.-K., Gibbens, R.J., Hancock, R.E., Towsley, D., Robust multipath routing in large wireless networks, in: *Proc. IEEE INFOCOM*, pp. 271–275, 2011.
22. Chen, P.-Y., Cheng, S.-M., Chen, K.-C., Smart attacks in smart grid communication networks. *IEEE Commun. Mag.*, 50, 8, 24–29, 2012.
23. Cho, J.-H. and Gao, J., Cyber war game in temporal networks. *PLoS One*, 11, 2, 24–29, 2016.
24. Cho, J.-H. and Moore, T., Percolation-based network adaptability under cascading failures, in: *Proc. IEEE INFOCOM*, pp. 2186–2194, 2018.
25. Cho, J.-H., Hurley, P.M., Xu, S., Metrics and measurement of trustworthy systems, in: *Proc. IEEE Mil. Commun. Conf. (MILCOM)*, pp. 1237–1242, 2016.
26. Cho, J.-H., Xu, S., Hurley, P.M., Mackay, M., Benjamin, T., Beaumont, M., STRAM: Measuring the trustworthiness of computer based systems. *ACM Comput. Surv.*, 128, 51, 6, 1–47, 2018.
27. Colbourn, C., Network resilience. *SIAM J. Alg. Discr. Meth.*, 8, 3, 404–409, 1987.
28. Erdös, P. and Rényi, A., On the evolution of random graphs. *Publ. Math. Inst. Hung. Acad. Sci.*, 5, 17–61, 1960.
29. Franceschetti, M., Dousse, O., Tse, D.N.C., Thiran, P., Closing the gap in the capacity of wireless networks via percolation theory. *IEEE Trans. Inf. Theory*, 53, 3, 1009–1018, 2007.
30. Duda, R.O., Hart, P.E., Stork, D.G., *Pattern Classification*, 2nd Edition, Wiley–Interscience, 2000. Available at: https://www.wiley.com/en-us/Pattern+Classification%2C+2nd+Edition-p-9780471056690
31. Fan, W., Miller, M., Stolfo, S.J., Lee, W., Chan, P.K., Using artificial anomalies to detect unknown and known network intrusions, in: *Proceedings of the IEEE International Conference on Data Mining*, IEEE Computer Society, pp. 123–130, 2001.
32. He, Z., Xu, X., Huang, J.Z., Deng, S., A frequent pattern discovery method for outlier detection. *Lecture Notes in Computer Science*. 3129, pp. 726–732, 2004.
33. Helmer, G., Wong, J., Honavar, V., Miller, L., Intelligent agents for intrusion detection, in: *Proceedings of IEEE Information Technology Conference*, pp. 121–124, 1998.
34. Hu, W., Liao, Y., Vemuri, V.R., Robust anomaly detection using support vector machines, in: *Proceedings of the International Conference on Machine Learning*, pp. 282–289, 2003.

35. Janakiram, D., Reddy, V., Kumar, A., Outlier detection in wireless sensor networks using Bayesian belief networks, in: *First International Conference on Communication System Software and Middleware*, pp. 1–6, 2006.
36. Muniyandi, A.P., Rajeswari, R., Rajaram, R., Network anomaly detection by cascading k-Means clustering and C4. 5 decision tree algorithms. *Proc. Eng.*, 30, 174–182, 2012.
37. Patel, J. and Panchal, K., Effective Intrusion Detection System using Data Mining Technique. *J. Emerg. Technol. Innov. Res.*, 2, 6, pp. 1869–1878, 2015.
38. Keerthi, S.S. and Gilbert, E.G., Convergence of a generalized SMO algorithm for SVM classifier design. *Mach. Learn.*, 46, 1-3, 351–360, 2002.
39. Collobert, R., Bengio, S., Bengio, Y., A parallel mixture of SVMs for very large-scale problems. *Neural Comput.*, 14, 5, 1105–1114, 2002.
40. Cataltepe, Z., Ekmekci, U., Cataltepe, T., Online feature selected semi-supervised decision trees for network intrusion detection. *Network Operations and Management Symposium (NOMS)*, pp. 1085–1088, 2016.
41. Muniyandi, A.P., Rajeswari, R., Rajaram, R., Network anomaly detection by cascading k-Means clustering and C4. 5 decision tree algorithms. *Proc. Eng.*, 30, 174–182, 2012.
42. Patel, J. and Panchal, K., Effective Intrusion Detection System using Data Mining Technique. *J. Emerg. Technol. Innov. Res.*, 2, 6, 2015.
43. Zareie, A. and Sakellariou, R., Minimizing the spread of misinformation in online social networks: A survey. *J. Netw. Comput. Appl.*, 186, 103094, 2021.
44. Jayachitra, S. and Galety, M.G., A smart management system for electric vehicle recharge using hybrid renewable energy. *J. Adv. Res. Dyn. Control Syst.*, 11, 1, 146–153, 2019.

8

Machine Learning Approach To Forecast the Word in Social Media

R. Vijaya Prakash

School of Computer Science & Artificial Intelligence, SR University, Warangal, India

Abstract

Forecasting is one of machine learning's most significant features. In machine learning, forecasting issues are classified as supervised learning algorithms. Label or target data process the given text to achieve a specific degree of accuracy or precision. The objective of time series analysis is to build models that best capture or explain the data to figure out what is driving a time series' fundamental causes. In this paper, words are forecasted on Twitter data with 1830 tweets. These tweets are interpreted as weighted documents using the term frequency and inverse document frequency (TF-IDF) algorithm. After obtaining the tweet's word frequency value, forecasting is done using the test data, and the findings accurately referred to 1303 slack word categories and 541 verb tweets as training data. Inactive users and active users were split into two categories when it came to word forecasting. The data were then processed using a MAPE calculation procedure that was set at 50% for inactive users and 20% for active users.

***Keywords*:** Time series, term frequency, forecasting, document frequency, text classification

8.1 Introduction

Text classification, prediction, and forecasting are considered a fundamental problem in information sciences. Text categorization, prediction, and forecasting are frequently used in the disciplines of information

Email: vijprak.r@gmail.com

Mohammad Gouse Galety, Chiai Al Atroshi, Bunil Kumar Balabantaray and Sachi Nandan Mohanty (eds.) Social Network Analysis: Theory and Applications, (133–148) © 2022 Scrivener Publishing LLC

sorting [1, 2] as an effective approach for text information organization and management. Deep learning is an efficient classification method [6, 7]. Furthermore, a growing number of researchers have employed frequently used neural networks to categorize and forecast text, such as the conditional random field [8] and the recurrent neural network (RNN) [3, 9].

Techniques that employ data to find a new framework in the computation of a specific involvement include classification, prediction, and forecasting [4, 5]. Neural networks have recently had a lot of success with text classification, and they are outperforming other methods. How to capture features for diverse units of text, such as phrases, sentences, and documents, is one of the issues in building a text categorization model. Recurrent neural network is particularly well suited to processing text of varying length because of its iterative structure [10, 11]. Recurrent neural network can better integrate information in some scenarios because it has a repeated network artefact that can be utilized to keep the data [12]. Long short-term memory (LSTM) [13, 14] and various versions [10] were created in a traditional RNN to address the problem of gradient bursting or disappearing. Long short-term memory performs well in the processing of natural language, living up to expectations.

Zuheros [15] in a paper explained the limitations of one of the deep learning methods in boosting performance and computing. To boost system performance and processing, they suggest a neural network design. Zhipeng Jiang [16] uses LSTM, which produces high accuracy but low processing performance. BiLSTM has also been employed by Beakcheol Jang [17] to increase text classification accuracy. Jason Wei [18] employed data augmentation techniques to increase text classification performance in another study. Using data augmentation techniques in the little data set, Francisco J Moreno [19] improved the accuracy value.

Term frequency (TF) and inverse document frequency (IDF) are combined in TF-IDF. Term frequency is the frequency of incidents of a term that appears in a document. Inverse document frequency reduces the weight of terms that appear frequently in the document collection and raises the weight of phrases that appear infrequently [20–22].

As a result, a categorization of the terms that occur on social media is required. Term frequency-inverse document frequency has not been identified in anticipating posted terms in social media in various earlier research. Text predicting on social media is required to determine the frequency value in social media postings. In social media word predictions and cooperation, the time series method is used. Better techniques are utilized to assess how effective the TF and IDF are at predicting texts using time series data.

Based on the abovementioned problems, we propose a new model, using TF-IDF algorithm with data augmentation techniques to improve the performance of clinical trials text classification.

8.2 Related Works

This section discusses research on this topic, with a focus on the use of data augmentation approaches to improve classification performance using current training data. Term frequency-inverse document frequency is extensively employed with classification of texts that have already exist, according to Chen [23] and Shouzhong & Minlie [24], and has been examined by numerous classification methods.

Using the KNN, TF-IDF techniques, Trstenjak, Mikac, and Donko [25] built a framework for text prediction. This framework highlighted the algorithm's benefits and drawbacks, as well as recommendations for further development on the same foundation. Rezaeian and Novikova [26] employed Naive Bayes in machine learning for text categorization based on conditional probability values. The Multinomial, Naive Bayes, Bernoulli, and Gaussian algorithm methods were used to compare the findings for statistical text representation.

Jamil and Hamzah [27] evaluated the execution of the Decision Tree (DT) and SVM in detecting feeling from Malay folk tale's data. Word-based feeling detection study was conducted using a collection of juvenile literature as a data set. The TF and IDF techniques were extracted from text materials and categorized using DT and SVM to identify emotions from Malay folk tales [24]. The goal is to lessen the occurrence of categorization errors by making it easier for humans to make mistakes, with a 91.25% accuracy rate, news may be classified using classification. However, for the advancement of online media, it is possible to categorize commonly used online media [28–30].

8.3 Methodology

Figure 8.1 depicts the overall structure of this investigation.

The following is an explanation of Figure 8.1:

1. Each tweet will be acquired from Twitter and the TF-IDF technique will be applied for frequency calculations.

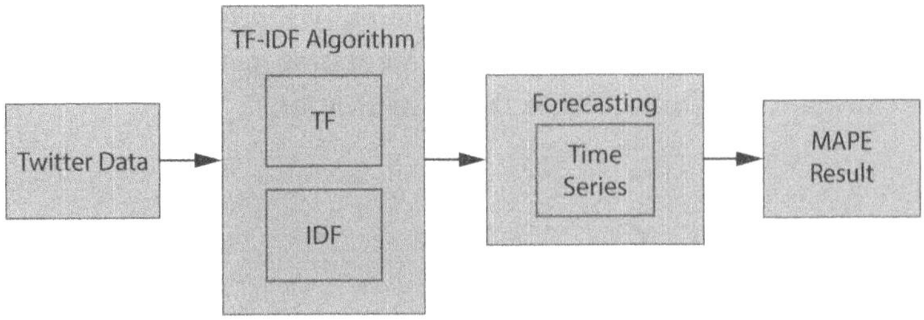

Figure 8.1 Overall methodology.

2. Text frequencies will be predicted using TF and IDF with a high frequency of recurrence using time series.
3. Get findings from time series categorization on social media word posts in the form of a comparison between expected and actual results.

8.3.1 TF-IDF Technique

The importance of each commonly exploited word is determined using the TF-IDF method [31]. This technique is well recognized for being cost-efficient, simple, and accurate [32]. The TF and IDF values for each token (word) in each document are calculated using this approach [33]. The TF and IDF methods are used to determine how often a document includes a text or a word.

We used the TF-IDF technique in this study, which was then combined with NBC [34]. As a result, the study's end conclusion is supposed to build a text categorization algorithm based on Twitter data. The effectiveness of the paper was measured by the number of tweets that occurred. First, find how frequently the word appeared in the document. As a result, the larger the frequency with which it occurs, the higher the value. There are various sorts of formulae that may be utilized in TF [35], including Binary TF, Pure TF, TF normalization, and TF logarithmic.

$$TF = \begin{cases} 1 + log_{10}(f_{t,d}), \text{ for } f_{t,d} < 0 \\ 0 \text{ for } f_{t,d} = 0 \end{cases} \tag{8.1}$$

The value of $f_{t,d}$ represents the frequency term t in the document d. Therefore, if a word or phrase appears five times in a text, the weight = 1 + log (5) = 1699 is determined. If the term is not included in the text, the weight is zero [36].

The IDF calculation measures how widely words in a set of texts are spread apart is the next step. In comparison to TF, the larger the value, the more often words are used. The greater the IDF number, the fewer words in the text display. To determine the quantity of the IDF value, use the formula below.

$$IDF_j = log\left(\frac{D}{df_j}\right) \tag{8.2}$$

where D is the total number of documents. The number of documents that include the term t_j is represented by df_j.

$$w_{ij} = tf_{ij} * idf_j \tag{8.3}$$

$$w_{ij} = tf_{ij} * log\left(\frac{D}{df_j}\right) \tag{8.4}$$

The number of words in a document is represented by w_{ij}. The number of times the term (t_j) appears in the document is represented by tf_{ij}. If D= df_j, then the result will be zero, regardless of the value of tf_{ij}, the IDF's result is log 1. As a result, on the IDF side, a value of 1 can be added, resulting in the following weight calculation:

$$w_{ij} = tf_{ij} * log\left(\frac{D}{df_j}\right) + 1 \tag{8.5}$$

8.3.2 Times Series

Time series refers to the collection and monitoring of data over a period [39] it uses seasonal data, cycles, trend data, and so on. Charts that go up or down over a period of 10 to 20 years show trend patterns. Seasonal data, on the other hand, fluctuate over a brief period, such as a year. This shows trends of ups and downs throughout time. This refers to the aspects that are not covered by the first three elements [40].

Several academics have used time series to model data [37]. Time series uses methods and models like ESM, SDM, VAM, SAM, and ARMAX models [38]. The time series approach uses previous data to forecast future data. The next value will be forecasted using the models that emerge. The R-squared model is not used to assess accuracy in time series. To see if the final equation is good or bad, utilize R-squared [41, 42].

In quantitative data analysis and forecasting, the trend method is a popular strategy. Because the model is used to forecast future data after searching for patterns in trend data like linear, quadratic, S curve, or exponential. The following models are used to predict the text with time series.

$$\text{Linear Model: Ypred} = a + bT + e \tag{8.6}$$

$$\text{Quadratic Model: Ypred} = a + bT2 + cT + e \tag{8.7}$$

$$\text{S curve Model: Ypred} = L/(1+\exp(a+b(T) + e) \tag{8.8}$$

$$\text{Exponential Model: Ypred} = a + eb.T \tag{8.9}$$

8.4 Results and Discussion

A general architectural study was created by categorizing a person's tweets using the Twitter data set. To determine the frequency of each word, the tweets were converted into documents. First, the TF value of each word was computed, with each word's weight set to one. Inverse document frequency is calculated using equation 8.2.

The outcome of the IDF computation is IDF_i. Weighting words are something to think about while looking for information from a diverse group of papers or tweets. A term in a document might be a single word or phrase that can be used to determine the document's context. Because each word in the document has a distinct amount of relevance, the term weight is used as an indication. Some of the procedures performed to determine the weight value using TF-IDF are represented in Table 8.1.

The TF-IDF calculation was performed using the term "good morning," which appears frequently, followed by the word happy, as shown in Table 8.1. However, the more words that occurred in the TF-IDF computation, the lower the frequency that appeared. To employ number processing, the frequency numbers were weighted. This figure was created in accordance with the study's objectives by computing time series, to anticipate remarks on data received from Twitter, and assessing the performance of a paper based on the twits that occurred. First, figure out how often words appear in a document. As a result, the greater the term's value, the higher its frequency of recurrence.

The training stage and the testing stage were stages involved in forecasting based on time series. The analysis in this study was done on samples in the form of Twitter twits. The terms that may occur in the document

Table 8.1 Twitter data set.

Term (t)	D_n	TF	IDF
Happy	0	42	0.46665582104149
Thank you	0	21	0.76768581670547
Anniversary	0	9	1.13566260200007
LOL	0	36	0.53360261067211
Love	0	13	0.97596175913256
Good morning	0	123	0

collection affected the social media users' habit of representing documents as much as feasible.

Twitter documents were employed in this study and were obtained with unstructured content. For categorization accuracy, structured papers were necessary, and they had to be comprehensible. To do the experiment analysis, two Twitter accounts are considered. One is inactive twitter account, and another is an active account. There are 1843 twits are collected from inactive account. Preprocessing is done prior to the prediction to assess the importance in training and testing texts because the data used are unstructured. It is difficult to understand the unstructured data, as it is represented in the form of xml and json objects. In the preprocessing stage, the data characteristics are identified to better analyze and apply the process.

Tokenization was the initial step in document processing to identify the tokens as some are used as delimiters. In general, the basic delimiters used in computer science are space, tab, and line characters. However, there are also other characters used as delimiters like #/?/@, and so on, depending on the usage. The tokenization procedure was then carried out because it is critical to detect the text patterns that will be anticipated, as well as the sorts of text that will be utilized for training, even though it had been processed using stop-words in the previous phase because of the irregularity of the text pattern discovered during identification. There was a lot of scepticism when it came to recognizing the text, and accuracy in observation was necessary, because the patterns were determined to be irregular in their content organization. To grasp the existing patterns in the text, the writer had to go over each document one by one throughout the identification process.

Training document label was manually constructed depending on the domain's selected category. In the document categorization process, label determination was used to give a reference. The data were divided into two categories: inactive users and active users, based on the results of document identification.

The usage of time series based on frequency and TF-IDF was used to forecast words on social media for inactive individuals. The term "happy" was coined to allow the findings of the TF IDF to be utilized as a weight in time series forecasting. Table 8.2 shows how forecasting was carried out in the system using the times series approach, with the results shown in Table 8.3.

The forecasting procedure may be computed, as can be shown in Table 8.3. The predicted results, on the other hand, were negative (–) or below zero. The "happy" artefact was created by utilizing the IDF TF of 0.466655821041497 for the prediction. The forecasting outcomes for inactive users may be seen through the graph, as illustrated in Figure 8.2.

The TF-IDF may be calculated using the frequency that commonly appears, such as the word "Love," shown in Table 8.4. However, the more words that occurred in the TF-IDF computation, the lower the frequency that appeared. To employ number processing, the frequency numbers were weighted. Using Twitter data with a weight of 1.222345, we computed time series to predict words. The twits that surfaced were then used to evaluate the document's performance.

Prior to projecting the data set utilized between October 2017 and August 2020, forecasting is performed later acquiring the TF-IDF value. Data were gathered from August 2017 to May 2020, with data testing in April 2020 and August 2020. Table 8.5 shows the active user's data on Twitter.

A data set of active Twitter users was presented in Table 8.5. Using the data from the TF-IDF computation, particularly the term "love," was learned through a time series technique. Figure 8.3 below shows the results of predicting the term "love." The blue data represent the training data, whereas the orange data are the testing data, as seen in Figure 8.3. Because the data "love" were retrieved from the TF-IDF computation, the term "love" was predicted. Using the MAPE formula, we may compare or assess the accuracy based on the testing results. However, you may examine the predicting results using the real data in Table 8.6 before completing the MAPE computation.

Table 8.6 shows that the predicting results for five testing data suggested four data with the identical outcomes, and the term "love" does not exist in the August 2020 predictions. The accuracy of this data was measured

Table 8.2 User data set is inactive.

Date	Data set	Date	Data set	Date	Data set
May 2010	0	December 2011	0	July 2013	0
June 2010	0	January 2012	0	August 2013	0
July 2010	5	February 2012	0	September 2013	0
August 2010	6	March 2012	0	October 2013	0
September 2010	0	April 2012	0	November 2013	0
October 2010	0	May 2012	5	December 2013	0
November 2010	0	June 2012	6	January 2014	0
December 2010	0	July 2012	0	February 2014	0
January 2011	0	August 2012	1	March 2014	4
February 2011	0	September 2012	0	April 2014	0
March 2011	0	October 2012	0	May 2014	0
April 2011	0	November 2012	0	June 2014	0
May 2011	0	December 2012	1	July 2014	0
June 2011	0	January 2013	0	August 2014	2
July 2011	0	February 2013	0	September 2014	4
August 2011	0	March 2013	0	October 2014	0
September 2011	18	April 2013	0	November 2014	2
October 2011	0	May 2013	0	December 2014	0
November 2011	0	June 2013	0	January 2015	0

using MAPE, which yielded a MAPE of 19.802%. This was the outcome of a modest MAPE, permitting word pattern predictions.

8.5 Conclusion

The researchers discovered that Twitter data might be anticipated, although each tweet was a beneficial action that revealed information about the person. This study employed a word prediction approach on social media,

Table 8.3 Forecasting the word "happy" in inactive users.

Date	Forecasting		Date	Forecasting
February 2015	0		February 2016	0
March 2015	2		March 2016	0
April 2015	2		April 2016	–1
May 2015	1		May 2016	2
June 2015	0		June 2016	1
July 2015	0		July 2016	1
August 2015	0		August 2016	0
September 2015	0		September 2016	0
October 2015	2		October 2016	–1
November 2015	1		November 2016	–1
December 2015	1		December 2016	2
January 2016	0		January 2017	1

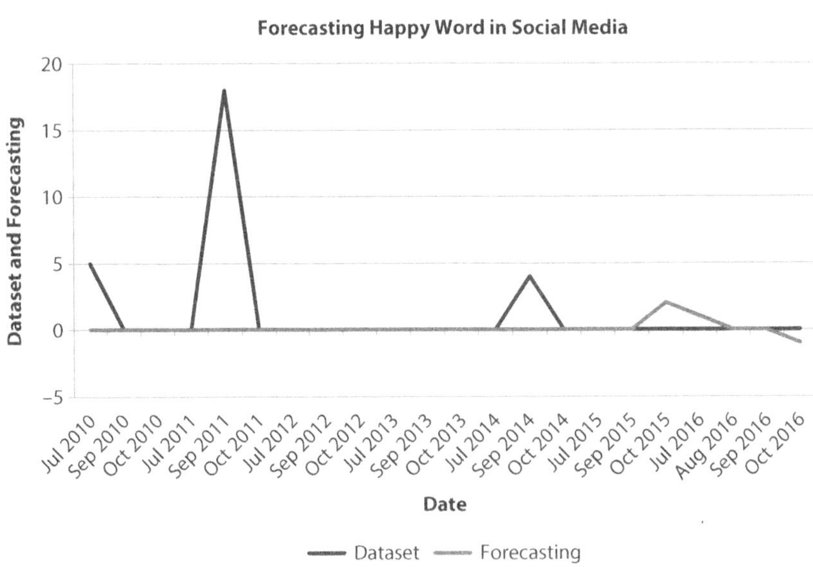

Figure 8.2 Forecasting happy word.

Table 8.4 Term data set active social media users.

Term (t)	Dn	DF	IDF
baby	0	35	1.3962
basic	0	25	1.7168
Love	0	23	1.2223
FTW	0	11	1.2148

Table 8.5 Data set of active user.

Date	Data set		Date	Data set
October 2017	0		March 2019	0
November 2017	0		April 2019	1
December 2017	0		May 2019	0
January 2018	0		June 2019	0
February 2018	0		July 2019	0
March 2018	0		August 2019	1
April 2018	0		September 2019	4
May 2018	1		October 2019	0
June 2018	2		November 2019	2
July 2018	0		December 2019	3
August 2018	0		January 2020	0
September 2018	1		February 2020	2
October 2018	0		March 2020	7
November 2018	3		April 2020	0
December 2018	1		May 2020	2
January 2019	0		June 2020	2
February 2019	2		July 2020	0
			August 2020	0

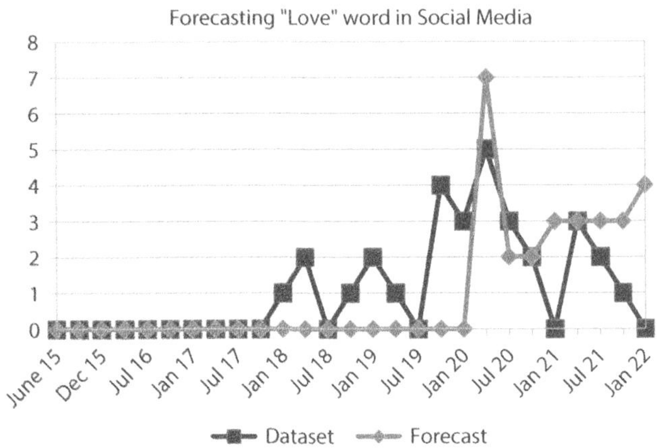

Figure 8.3 Forecasting "love" word.

Table 8.6 Error and MAPE.

Date	Actual	Forecasting
Apr 2020	5	7
May 2020	2	2
June 2020	2	2
July 2020	3	2
August 2020	0	1

starting with the TF-IDF calculation. According to this, TF-IDF values with frequent occurrences have a lower frequency value, whereas TF-IDF values with fewer occurrences have a higher frequency value. Following the TF-IDF calculations, forecasting was carried out using a time series method and a system that divided word forecasting into two categories: idle users and active users. The test data for idle users contain 1734 tweets with 1203 slack keyword categories and 531 twits, whereas the test data for active users has 584 tweets with 60,613 words. The results were obtained utilizing a MAPE calculation technique that included a 50% idle user rate and a 19.8% active user rate.

References

1. Zhang, J., Liu, F., Xu, W., Yu, H., Feature fusion text classification model combining CNN and BiGRU with multi-attention mechanism. *Future Internet*, 11, 11, 237–261, Nov. 2019.
2. Pavitra, R. and Kalaivaani, P.C.D., Weakly supervised sentiment analysis using joint sentiment topic detection with bigrams. *2nd Int. Conf. Electron. Commun. Syst. ICECS 2015*, pp. 889–893, 2015.
3. Liao, S., Wang, J., Yu, R., Sato, K., Cheng, Z., ScienceDirect CNN for situations understanding based on sentiment analysis of twitter data. *Proc. Comput. Sci.*, 111, 2015, 376–381, 2017.
4. Xie, S., Wang, G., Lin, S. and Yu, P.S., 2012, August. Review spam detection via temporal pattern discovery. In Proceedings of the 18th ACM SIGKDD international conference on Knowledge discovery and data mining, pp. 823–831, 2012.
5. Wang, A.H., Don't follow me: Spam detection in twitter, *In 2010 international conference on security and cryptography (SECRYPT 2010)*, pp. 1–10, 2010.
6. Jasmir, J., Nurmaini, S., Malik, R.F. and Zaenal, D., Text Classification of Cancer Clinical Trials Documents Using Deep Neural Network and Fine Grained Document Clustering, *Architecture*, 10, pp. 396–404, 2020.
7. Jasmir, *et al.*, Breast Cancer Classification Using Deep Learning. *Proc. 2018 Int. Conf. Electr. Eng. Comput. Sci. ICECOS 2018*, vol. 17, pp. 237–242, 2019.
8. Jasmir, J., Nurmaini, S., Malik, R.F. and Tutuko, B., Bigram feature extraction and conditional random fields model to improve text classification clinical trial document, *Telkomnika*, 19, 3, pp. 886–892, 2021.
9. Y. Li, X. Wang, and P. Xu, Chinese Text Classification Model Based on Deep Learning, *Future Internet*, 10, 11, pp. 113–125, 2018.
10. Cho, K. *et al.*, Learning phrase representations using RNN encoder-decoder for statistical machine translation. *EMNLP 2014 - 2014 Conf. Empir. Methods Nat. Lang. Process. Proc. Conf.*, pp. 1724–1734, 2014.
11. Darmawahyuni, A., Nurmaini, S., Caesarendra, W., Bhayyu, V. and Rachmatullah, M.N., Deep learning with a recurrent network structure in the sequence modeling of imbalanced data for ECG-rhythm classifier, *Algorithms*, 12, 6, pp. 118–130, 2019.
12. Sari, W.K., Rini, D.P., Malik, R.F. and Azhar, I.S.B., Sequential Models for Text Classification Using Recurrent Neural Network, *In Sriwijaya International Conference on Information Technology and Its Applications* (SICONIAN 2019), pp. 333–340, 2020.
13. Gulli, A. and Pal, S., Long Short-Term Memory - LSTM, in: *Deep Learn. with Keras*, pp. 187–195, 2017.
14. Darmawahyuni, A., Nurmaini, S., Sukemi, Deep Learning with Long Short-Term Memory for Enhancement Myocardial Infarction Classification. *Proc.*

2019 6th Int. Conf. Instrumentation, Control. Autom. ICA 2019, no. August 2019, pp. 19–23, 2019.
15. Zuheros, C., Tabik, S., Valdivia, A., Martínez-Cámara, E. and Herrera, F., Deep recurrent neural network for geographical entities disambiguation on social media data, *Knowledge-Based Systems*, 173, pp. 117–127, 2019.
16. Liu, Z., Tang, B., Wang, X., Chen, Q., De-identification of clinical notes via recurrent neural network and conditional random field. *J. Biomed. Inform.*, 75, S34–S42, 2017.
17. Jang, B., Kim, M., Harerimana, G., Kang, S.U., Kim, J.W., Bi-LSTM model to increase accuracy in text classification: Combining word2vec CNN and attention mechanism. *Appl. Sci.*, 10, 17, 5841–5855, 2020.
18. Wei, J. and Zou, K., EDA: Easy data augmentation techniques for boosting performance on text classification tasks. *EMNLP-IJCNLP 2019 - 2019 Conf. Empir. Methods Nat. Lang. Process. 9th Int. Jt. Conf. Nat. Lang. Process. Proc. Conf.*, pp. 6382–6388, 2020.
19. Moreno-Barea, F.J., Jerez, J.M., Franco, L., Improving classification accuracy using data augmentation on small data sets. *Expert Syst. Appl.*, 161, 113696–113710, 2020.
20. Rahmah, A., Santoso, H.B., Hasibuan, Z.A., Exploring Technology-Enhanced Learning Key Terms using TF-IDF Weighting. *Proc. 2019 4th Int. Conf. Informatics Comput. ICIC 2019*, 2019.
21. Li, G. and Li, J., Research on Sentiment Classification for Tang Poetry based on TF-IDF and FP-Growth, in: *2018 IEEE 3rd Advanced Information Technology, Electronic and Automation Control Conference (IAEAC)*, pp. 630–634, 2018.
22. Arroyo-Fernández, I., Méndez-Cruz, C.F., Sierra, G., Torres-Moreno, J.M., Sidorov, G., Unsupervised sentence representations as word information series: Revisiting TF–IDF. *Comput. Speech Lang.*, 56, 107–129, 2019.
23. Chen, K., Zhang, Z., Long, J., Zhang, H., Turning from TF-IDF to TF-IGM for term weighting in text classification. *Expert Syst. Appl.*, 66, 1339–1351, 2016.
24. Shouzhong, T. and Minlie, H., Mining microblog user interests based on TextRank with TF-IDF factor. *J. China Univ. Posts Telecommun.*, 23, 5, 40–46, 2016.
25. Trstenjak, B., Mikac, S., Donko, D., KNN with TF-IDF based framework for text categorization. *Proc. Eng.*, 69, 1356–1364, 2014.
26. Rezaeian, N. and Novikova, G., Persian text classification using naive bayes algorithms and support vector machine algorithm. *Indones. J. Electr. Eng. Inform.*, 8, 1, 178–188, 2020.
27. Saad, M.M., Jamil, N., Hamzah, R., Evaluation of support vector machine and decision tree for emotion recognition of Malay folklores. *Bull. Electr. Eng. Inform.*, 7, 3, 479–486, 2018.

28. Seo, D.B. and Ray, S., Habit and addiction in the use of social networking sites: Their nature, antecedents, and consequences. *Comput. Hum. Behav.*, 99, May, 109–125, 2019.
29. Dandannavar, P.S., Mangalwede, S.R., Kulkarni, P.M., Social Media Text - A Source for Personality Prediction. *Proc. Int. Conf. Comput. Tech. Electron. Mech. Syst. CTEMS 2018*, pp. 62–65, 2018.
30. Devi, K.S., Gouthami, E., Lakshmi, V.V., Role of Social Media in Teaching – Learning Process. *J. Emerg. Technol. Innov. Res.*, 6, 1, 96–103, 2019.
31. Liu, Q., Shao, Z., Fan, W., The impact of users' sense of belonging on social media habit formation: Empirical evidence from social networking and microblogging websites in China. *Int. J. Inf. Manage.*, 43, 13, 209–223, 2018.
32. Yuan, M. and Zou, C., Text Keyword Extraction Based on Meta-Learning Strategy. *Int. Conf. Big Data Artif. Intell. BDAI 2018*, pp. 78–81, 2018.
33. Rahmat, M.A., Indrabayu, Areni, I.S., Hoax web detection for news in Bahasa using support vector machine. *2019 Int. Conf. Inf. Commun. Technol. ICOIACT 2019*, pp. 332–336, 2019.
34. Langseth, H. and Nielsen, T.D., Classification using Hierarchical Naïve Bayes models. *Mach. Learn.*, 63, 2, 135–159, 2006.
35. Mohammedid, M. and Omar, N., Question classification based on Bloom's taxonomy cognitive domain using modified TF-IDF and word2vec. *PLoS One*, 15, 3, 1–15, 2020.
36. Jalilifard, A., Caridá, V.F., Mansano, A.F., Cristo, R.S. and da Fonseca, F.P.C., Semantic sensitive TF-IDF to determine word relevance in documents, *In Advances in Computing and Network Communications*, pp. 327–337. Springer, Singapore, 2021
37. Esling, P. and Agon, C., Time-series data mining. *BodyNets Int. Conf. Body Area Networks*, 2012.
38. Mishra, N., Soni, H.K., Sharma, S., Upadhyay, A.K., A comprehensive survey of data mining techniques on time series data for rainfall prediction. *J. ICT Res. Appl.*, 11, 2, 167–183, 2017.
39. Zhu, Z. *et al.*, Continuous monitoring of land disturbance based on Landsat time series. *Remote Sens. Environ.*, 238, November 2018, 111116, 2020.
40. Wang, Y. *et al.*, Time series analysis of temporal trends in hemorrhagic fever with renal syndrome morbidity rate in China from 2005 to 2019. *Sci. Rep.*, 10, 1, 9609, 2020.
41. Golyandina, N., Particularities and commonalities of singular spectrum analysis as a method of time series analysis and signal processing. *Wiley Interdiscip. Rev. Comput. Stat.*, 12, 4, 1–46, 2020.
42. Saravanan, S., Hailu, M., Gouse, G.M., Lavanya, M., Vijaysai, R., Design and analysis of low-transition address generator, in: *Lecture Notes of the Institute for Computer Sciences, Social-Informatics and Telecommunications Engineering*, LNICST, vol. 274, pp. 239–247, 2019.

9

Sentiment Analysis-Based Extraction of Real-Time Social Media Information From Twitter Using Natural Language Processing

Madhuri Thimmapuram[1*], Devasish Pal[2] and Gouse Baig Mohammad[1]

[1]Department of Computer Science and Engineering, Vardhaman College of Engineering, Shamshabad, Hyderabad, India
[2]Department of Information Technology, Muffakham Jah College of Engineering & Technology, Hyderabad, India

Abstract
Social networking has rapidly expanded to include millions of individuals throughout the world. It allows users to develop and share its material in various types of information, personal text, image, audio, and videos through this kind of electronic communication through social networking platforms. Thus, social computing has become the new subject of study and development, which covers a wide range of concerns, including Internet semantics, artificial intelligence, linguistic processing, network analysis, and big data analytics. In the previous few years, we have transformed and altered our online social network approach with individuals, groups, and communities (Facebook, Twitter, YouTube, Flickr, MySpace, LinkedIn, Metacafe, Video, and so on). We are developing in this study a program that analyzes the nature of tweets on a specific idea. The main goal is to analyze polarity in noisy Twitter streams. This work reports on the conception of a data analysis, which extracts many tweets. Results divide users into positive and negative perceptions via tweets. The user can enter a keyword and learn the nature of this on the basis of the latest tweets containing the keyword input. Each tweet is classified on the basis of a favorable or bad feeling. Data are collected regarding film reviews from the IMDB website. The machine learning algorithm Naive Bayes was utilized. Different test methods were used to test the result of this model. In addition, our algorithm is quite effective on mining

**Corresponding author*: munny.vsp@gmail.com

Mohammad Gouse Galety, Chiai Al Atroshi, Bunil Kumar Balabantaray and Sachi Nandan Mohanty (eds.) Social Network Analysis: Theory and Applications, (149–174) © 2022 Scrivener Publishing LLC

sentences directly taken from Twitter. The accuracy was 92.50% with good generalization capabilities and good speed of execution.

***Keywords*:** Natural language processing, Naïve Bayes, social network, microblogging, sentiment analysis

9.1 Introduction

On social media platforms, there are numerous ways to interact. One of the main things is through text messages. In the past 25 years, conventional media such as printed news and articles have been used in natural language processing (NLP).

The NLP generally allows computers the use of computer expertise, artificial intelligence, and linguistics to derive meaning from their natural language input. More details on social media sites and their characteristics and shared content types are given in Table 9.1 [1].

Social media statistics for January 2014 show that over 1 billion active users have been on Facebook with a total annual turnover of more than 200 million. Statista, the world's largest directory of statistics, announced a social networking ranking based on the number of active members. The ranking shows that Qzone was second with more than 600 million users as depicted in Figure 9.1. Google+, LinkedIn, and Twitter completed the top 5 with 300 million, 259 million, and 232 million active users.

In social media platforms, there are numerous ways to interact. One of the main things is through text messages. In the past 25 years, conventional media such as printed news and articles have been used in natural language processing (NLP). The NLP generally allows computers the use of computer expertise, artificial intelligence, and linguistics to derive meaning from their natural language input.

For social media text, the NLP is a new field of research that needs to be adapted to this type of texts or to build new methods that fit for extracting information and other social media work. For various reasons (for instance, informal character, new language, abbreviations, etc.), the "traditional" NLP is not enough for social media messages.

There is a social network consisting of several actors (for example, persons and organizations) and binary relationships (such as connections or interactions). The aim is to construct a social group structure from a social network perspective to understand the impact on other parts of this structure and how structures evolve over time. The Semantic Social Media Analysis (SASM) processes text communications semantically and meta-data for the development of intelligent applications based on social media data.

Table 9.1 The characteristics of social media platforms.

Type	Characteristics	Examples
Social networks	A social networking website allows the user to build a web page and connect with a friend or other acquaintance to share user-generated content.	Myspace, Facebook, LinkedIn, Meetup, Google Plus+
Blogs and blog comments	A blog is an online journal where the blogger can create the content and display it in reverse chronological order. Blogs are generally maintained by a person or a community. Blog comments are posts by users attached to blogs or online newspaper posts.	Huffington Post, Business Insider, Engadget, and online journals
Microblogs	A microblog is similar to a blog but has a limited content.	Twitter, Tumblr, Sina Weibo, Plurk
Forums	An online forum is a place for members to discuss a topic by posting messages.	Online Discussion Communities, phpBB Developer Forum, Raising Children Forum
Social bookmarks	Services that allow users to save, organize, and search links to various websites, and to share their bookmarks of web pages.	Delicious, Pinterest, Google Bookmarks
Wikis	Lese websites allow people to collaborate and add content or edit the information on a community-based database.	Wikipedia, Wikitravel, Wikihow
Social news	Social news encourages their community to submit news stories, or to vote on the content and share it.	Digg, Slashdot, Reddit
Media sharing	A website that enables users to capture videos and pictures or upload and share with others.	YouTube, Flickr, Snapchat, Instagram, Vine

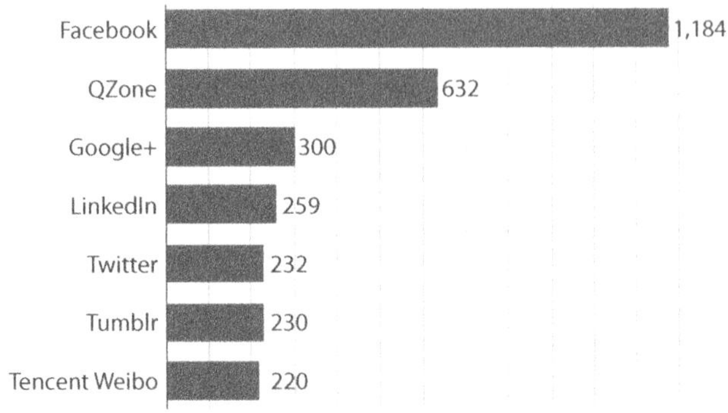

Figure 9.1 Statista presented a ranking of social networks based on the number of active users as of January 2021 (in millions).

To prevent users from behavior, or to extract other kinds of information, SASM helps to design automated tools and algorithm to monitor, gather, and analyze massive volumes of data acquired from the social media. Where the volume of data is particularly vast, "big data" processing techniques need to be utilized (for example, online algorithms that do not have to store all data to update the models on the basis of input data).

Publicly accessible texts are collected through blogs and microblogs, online forums, FAQs, chat, podcasts, online games, tags, ratings, and commentaries by social media sources. The information is available from the publicly accessible publication. Social media texts have various qualities, because the nature of the social talks, published in Echtzeit, differs from traditional texts. It is vital to detect groupings of topical chats as well as emotions, rumors, and incentives for applications. Also, significant information can be added by defining the places indicated in messages or users' locations. The text is unstructured and is available in multiple formats and written in many languages and styles by many people. Also, on social networking sites, such as Facebook and Twitter, typographic errors and chat lingo have become more common. The authors are not professional writers, and their posts may be seen on several social media platforms on the web in many locations.

Tracking and analyzing the rich and ongoing flow of user-generated content may produce significant, unprecedented information not available through traditional media outlets. The emergence of the science of Big Data Analytics is derived from the seminal analysis of social media, machine education, data mining, information retrieval, and processing of natural languages [2].

Figure 9.2 presents the framework for the semanticization of social media. The first stage is to identify issues and opportunities for collecting

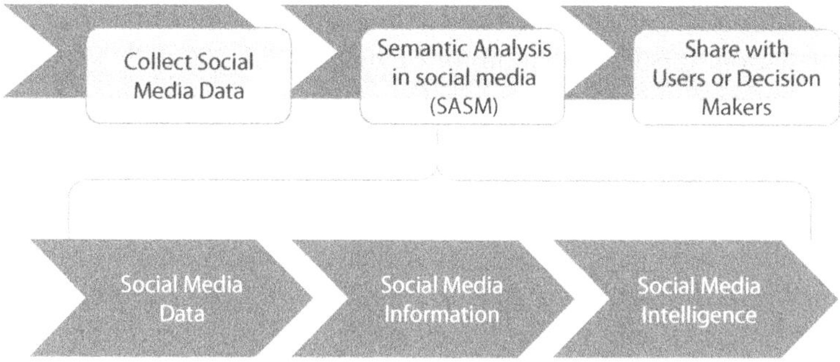

Figure 9.2 The semantic analysis framework in social media that transforms data into intelligence through NLP technologies.

social network data. The information can be kept in text (large, complex, or text files can retain the data), the online collection of data can be processed in real time or retrospectively for specific purposes to be processed. The next phase is the SASM pipeline, comprised of customized NLP tools that analyze and process social media. Social media information consists of vast, noisy, and unstructured data sets. By social information and expertise, SASM converts social media data into relevant and understandable communications. Semantic Social Media Analysis then analyzes the information on social media to generate information on the social media. Social media intelligence may be shared or given to decision makers for awareness raising, communication, planification, and resolution of issues. Data visualization methods can be used to present the analyzed data via SASM.

The online media has become an important tool for people to communicate their views [1, 2], and there is plenty of information available with the social media. The polarity of opinions can be discovered by studying the text of the viewpoint with sentimental analysis, such as positives, negative, or neutral [3]. The analysis of feelings was helpful to the corporations for their products [5], foresight results of elections [6], and film review opinions [4]. The analyses of sentiment have been helpful to companies. For companies that take future decisions, the knowledge collected from sentiment analysis is useful [2, 5]. In sentiment analysis, many conventional approaches employ a word bag [7].

9.1.1 Applications for Social Media

Automatic processing of social media needs the development of relevant applicable research processes, such as information collecting, automated categorization, clustering, data collection indexing, and machine statistical translation. It is hard to monitor or analyze any manually useful information because of the volume of information from social media and the enormous rate of new content production. In many situations, the volume of the data is too large for a decision maker to effectively evaluate human data in real time and analyze it.

One of the keys used in SASM is social media surveillance. Media surveillance is traditionally designed to monitor and track the output of hard copying, Internet, and media broadcasting that may be conducted on a range of grounds, including politics, business, and science. A major source for open intelligence is the vast amount of information offered through social media networks. Social media allow direct touch with the audience aim. In contrast to traditional journalism, authors' opinions and sentiments provide social media data with an additional dimension. It is also difficult to analyze social media documents in different sizes, such as a mix of various tweets and blogs and content variability.

The search queries in social media cover various dimensions, including space and time, for real-time event search or event detection. Certain NLP approaches, such as the collection of information and the summary of social data as distinct documents from various sources, are crucial for the search for events and for identification of the relevant data in this situation.

A semantical examination of the meaning of talks in social networks across days and weeks for a group of topical debates and events exposes the problems of cross-language NLP activities. Social media-related NLP techniques that can extract an analyst's preference integration data have also led to computer-language domain-based applications.

9.1.2 Social Media Data Challenges

The information provided in social media is highly dynamic and includes interactions among numerous users, such as online forums, blogs, and Twitter updates. In informal situations, there are a large volume of text generated continuously by users.

Therefore, the typical NLP approaches used in social media text face difficulties because of un-standard orthography, noise, and restricted collection of automated grouping and classification characteristics. Social media is essential because everybody has become a prospective author with the usage of social networks, therefore, the language is now closer to the user than to any prescriptive standards [8]. In an informal and conversational style, blogs, tweets, and updates on the status are often produced in no more "stream of consciousness" than the carefully designed and thoroughly edited work that traditional printed media may anticipate. This informal style of social media communications offers new obstacles for artificial language processing on all levels.

Several problems on the surface present problems for basic NLP methods for traditional data. Inconsistent (or nonexistent) punching and shaping of sentences can make it rather difficult to recognize sentence borders—sometimes even for human readers, as in the following tweet: "#qcpoli had a hearty laugh with the @jflisee #notrehome debate crowd today, that the planned reaction?" Speech tactics and partial speaker complicate among other duties, emotics, erroneous, or nonstandard spelling, or abbreviated abbreviations. New modifications, such as the letters repeating ("heyyyyyy"), which are different than conventional mistakes must be considered in traditional techniques. Grammar, or frequently, the lack of it, is another difficulty for any syntactic analysis of social media communications, in which pieces can be as prevalent as genuine complete phrases, and choices can appear to occur at random between "there," "they are," "they are," and "their."

Social media is also significantly more noisy than conventional printed media. Like much everything in the Internet, spam, advertisements, and other unwanted or distracting content are plaguing social network. Most factual, authorized stuff in social media can be regarded irrelevant even if such noise is ignored for most information requests. In a research assessing the perceived worth of tweets, the author in [9] shows this. More than 40,000 tweets have been collected by followers, in which 36% of tweets are assessed as "worthy," whereas 25% are classified as "not worth reading." The lowest-rate tweets (e.g., "Hullo Twitter!") were known as presence maintenance posts. It is necessary to preprocess spam and any other relevant content or models, which are more suited to noise in any language processing effort aimed at social media.

The NLP techniques are particularly concerned with certain features of social media texts. The characteristics of the media and the manner it is employed might have an impact profoundly on the approach to summarize success. The 140-character limit on Twitter messages, for example, imposes an impoverished text on each of the tweets compared with more typical papers. However, duplication is a concern due partly to the practice of retweeting posts over several tweets. Automatically building summary postings on Twitter trends in their experiments with data mining techniques by the author in [10] notes that the redundancy of content represents a big issue with microblog resuming.

The management and processing of Twitter's ever vast volume of data demand highly scalable and efficient technologies (especially for real-time event detection). The design and use of Twitter are also challenging. These pertain mostly to the short-sighted usage of informal, irregular, and abbreviated words (dynamically changing). The great quantity of orthographing and grammatical faults. Such data sparsity, context, and lexical diversity make the traditional tweets less appropriate for text analysis algorithms. Moreover, various events might benefit from a different popularity among the users: their content, numbers of messages and participants, periods, inherent structure, and causal connections can differ considerably [11].

Subjectivity is an ever-present feature in all types of social media. Although conventional news copy may attempt to give the factual facts an objective, balanced report, social media text is significantly more subjective and fuller of opinion. In the semantine analysis of social texts, the necessity for or not the final information lies directly in opinion mining and sentiment analysis.

As the Internet grows, it becomes wider to the public. In sharing encapsulated news and hot subjects around the globe, social media, and microblogging platforms like Twitter, Facebook, Tumblr dominate at a rapid rate. If many users contribute their opinions and judgments, a subject or news becomes trending and thus a vital source of Internet perception for this particular issue. Themes, such as awards, films, and popular electoral personalities, are frequently used to create awareness or promote political camps, product approvals, and entertainment. Large corporations and organizations to improve marketing tactics make advantage of people's opinion on these platforms. One such example is to leak photos of the next iPhone to inflate people's feelings and advertise the product before it is released. Therefore, the potential to uncover and analyze significant patterns for business-driven applications from innumerable social media data is enormous. Emotions are predicted while analyzing feelings in a word, sentence, or corpus. It seeks to understand Internet beliefs, attitudes, and emotions. The objective is to gain an insight into or to get to know the general audience behind specific topics. It is precisely a paradigm that classifies conversations as positive, bad, or neutral. Many individuals use social networking websites to keep up with news and happenings. Services like Twitter, Facebook, Instagram, and Google+ provide voice opinions to individuals. For example, you submit your reviews online instantaneously when you see a film and then start commenting on the performance skill of the movie. This information offers a basis for individuals to evaluate the performance not just of film but of other items and to determine whether it is successful. These websites may be utilized with this sort of information in marketing and social studies. Sentiment analysis, therefore, has extensive applications and involves mining of emotions, polarity, classification, and analysis of influence. Twitter is a website run by tweets with a fixed number of 280 characters. Therefore, the limit of the character compels the usage of hashtags to classify material. Almost 6,500 tweets are published per second, leading to around 561,6 million tweets every day. These tweets are typically bruising and reflect multiple topics of changing feelings. Twitter sentiment analysis is the use of NLP to extract and detect the feeling substance.

In this work, we will look at any sporting event, movie trends in one time and analyze the public opinion on the social media network Twitter using real time data. By using the tweets, we try to forecast the mass views represented in the tweets or make a choice. Twitter allows companies to access broader audiences with more than 321 million consumers active and to send an average of 500 million tweets per day without an intermediary. On the other hand, brand detection of unfavorable information is more difficult, and if it spreads virally, you may wind up with an unforeseen PR catastrophe. This is one of the causes for social listening. Social media monitoring and feedback has become a critical social media marketing procedure. Twitter monitoring enables firms to comprehend and keep abreast of what their brands and competitors are saying and learn about emerging trends in the market. Are users commenting about a product either positive or negative? Well, that is exactly what sentiment analysis determines.

The rest of the chapter is planned as follows: in section 9.2, literature survey; section 9.3, we give information about the methodology, implementations, performances, and experimental at that point execution and investigation of this work; section 9.4 and section 9.5 give the detailed information about the conclusion with recommendations for future work.

9.2 Literature Survey

Although substantial study in the subject of sentimental analysis has been carried out over a few decades, the transition to social networks and blogs can be traced back to the early part of this decade. The enormous material available on social networking platforms are popular spots to convey your ideas to scientists to experiment with model sentiment analyses that can make them feel. One of the earliest efforts on the development of Twitter sentiment can be seen in the [12] that tried to learn machine algorithms like "Naive Bayes," "maximum entropy," and "SVM" on tweet data. The work utilized Emoticon to set up classificator training sets and proved the efficiency and accuracy of Twitter data with machine learning techniques at 82.2% for unigrams. The author in [13] helped to extract the Twitter feeling by constructing a massive corpus. They streamed and categorized data using the Twitter API. This data collection was used until early 2013, when the majority of studies on the extraction of the Twitter feelings had to be removed according to the Twitter authority's notification. A two-stage classification approach [3], which classifies communications as subjective and objective, followed by subjective tweets as positive and negative, which filters a Neutral Tweet before a Binary Classification were made. They recommended an extract of the metadata in the tweets and unigrams for the small data size of the training models in a polarity classification. In an automated model [4], the latent properties of users were explored to extract features that could detect and, consequently, categorize qualities, such as gender, age, geographical origin, actual policies, and so on.

The author in [5] conducted a study of a large collected body in the language and showed how a sentiment classifier may be built using the collected body as data for training. They have tried with a Naive Bayes multinomial classification and achieved the best results for bigram compared with unigrams, trigrams, and POS. Researchers also indicated a desire for investigating the association between Twitter sentiment and the facts [6], in 2008 to 2009, which collected more than 1,000 billion tweets on three political themes. It was the presence or absence of emotive phrases that accumulated characteristics. The study utilized correlations to indicate the ability to imitate traditional polling results in text streams. The results vary between the data sets with up to 80% correlations. The author in [7], which was about analyzing Twitter's text content with the help of mood monitoring technologies, particularly Opinion Finder and Google's Mood Profile States (GPOMS), compared the public response with the presidential election and Thanks for Day in 2008. Apple's results were cross validated. A similar work [8] tried to retrieve tweets on politicians or political parties. More than one lakh tweets were collected, and sentiment analysis results were compared with the actual.

Tools, such as LIWC and Opinion Finder for text analysis, were utilized to extract various emotions from speakers. Another similar piece [9] in that genre was an attempt at sentimental analysis of the tweets that were published in Twitter using a psychometric instrument to derive six moods from the aggregates Twitter material. In [10], the POS-relevant previously polarized features were introduced by Sentiwordnet, and a tree kernel was developed that exceeded the unigram model. The useful language of Twitter messages has been examined [11] on the Twitter corpus in Edinburgh, improving its findings marginally beyond conventional baselines. Their results have been marginally enhanced. In [12], which was investigated with data streaming via the Twitter API, a hybrid strategy to combine dictionary and corpus approaches was tried. The corpus technique has been utilized to identify the semantic orientation of adjectives and to handle the verbs and adverbs using the dictionary method. A general lexicon-based technique was suggested by [13] to infer customer's sentiment from tweets. The automatic sentiment analysis approaches that have been used have been reviewed [14], the textual properties of the social media messages are researched in the context of the development of methods of analysis of sentiment, and Twitter messages have been recommended for automatic sentiment analysis. A Barack Obama-McCain Debate (OMD), a generic data set and a Stanford Twitter Sentiment (STS), and a healthcare data set for semantic analysis of Twitter data, mostly using a pre-determined versatility [15], were used.

The author in [16] collected helpful information from the Twitter website and performed an efficient sentimental analysis of smart phone conflict tweets. For the prediction of the age of the use, it utilizes an efficient scoring system. A well-trained Naive Bayes Classifier is predicted for the user sex. The Tweet is labeled by the Sentiment Classifier Model with a feeling that allows to fully analyze data using multiple consumer criteria, such as location, sex, and age. A feel-good analysis technique was presented, particularly for Chinese microblogs [17] that included parallelization to minimize time complexity and optimize data structure. Recently, an innovative real-time system has been presented for the implementation of sentient analysis on social media figures [18], which implements and compares the SVM and Naive Bayes algorithm. A new approach was developed in [19], which performs a detailed investigation on emotional indications and interrelationships to deciphering emotional signals for unchecked sentimental analysis.

One of the first research studies on tweet polarity classification was [20]. The authors have performed a supervised classification analysis on tweets in English utilizing emoticons (e.g., ":"), ":(" etc ...). Using this technique, a corpus of good tweets, positive emoticons ":)" and bad emoticons tweets ":(") was generated They subsequently utilize various supervised algorithms and different sets of features (SVM, Naive Bayes, and Maximum Entropy) and find that the mere usage of unigrams leads to good results, but that the combination with unigrams and bigrams may marginally improve this strategy. In the same line of thinking [4], a corpus of tweets was also produced to analyze sentiment by choosing good and negative tweets based on particular emoticons. They then use Naive Bayes with unigrams and part of speech tags to compare several monitored techniques with n-grams and to reach the highest outcomes.

Another way to analyze your feelings in a tweet is [21]. In this regard, authors use a hybrid strategy, integrating supervised education with the knowledge they draw from the DAL sentiment dictionary [22]. Your pre-processing stage involves removing retweets, translating abbreviations into original phrases, removing links, a method of tokenization, and part-of-speech tagging. They use a variety of monitored learning algorithms to categories tweets into positive and negative utilizing SVM N-Gram features and Partial Tree Kernels syntactic features, in combination with knowledge about the multi-tweet word polarity. The authors find that those which match sentimental terms are the most essential qualities. Finally, [23] categories in tweets the feeling that has already been conveyed on "targets." You add the tweet context information to your text (e.g., the event that it is related to). They then work at SVM and General Inquirer and do a classification in three directions (positive, negative, and neutral).

To classify Twitter's message level sentiment [20], a comprehensive education approach has been developed. A characteristic representation that mixes state-of-the-art features and the emotive word is made in ten million tweets collected in the form of happy and negative emotions and without manual comments. Most work attempted with machine learning or generic lexicon to develop a classifier model. POS characteristics have been shown not to contribute to the classification of sentiments. In extracting tweets, the analysis of previous studies does not demonstrate a domain-specific technique. Domain particular ways to improve the polarity of the lexicons were attempted, and significant results were reported on formal text patterns [21, 22]. However, a pure lexicon-based strategy does not have to function with a restricted text size of microblogs. We must try to extract the feelings in the tweet to complement standard models, such as punctuations, emojis, hashtags, and so on.

The management and processing of Twitter's ever vast volume of data demand highly scalable and efficient technologies (especially for real-time event detection). The design and use of Twitter are also challenging. These pertain mostly to the short-sighted usage of informal, irregular, and abbreviated words (dynamically changing). The large amount of grammatical and spelling defects, the scarcity of information, the lack of context and the diversity of the lexicon render the usual methods of text analysis less appropriate for tweets [24]. Different events can also be very popular among users; the content, number of messages and participants, periods, structure, and causative links can vary greatly [25]. Subjectivity is an ever-present feature in all types of social media. While conventional news copy may attempt to give the factual facts an objective, balanced report, social media text is significantly more subjective and full of opinion. In the semitone analysis of social texts, the necessity for or not the final information lies directly in opinion mining and sentiment analysis.

In social media, drifting is far more essential in terms of the conversational tone of social texts and the continuous flow of social media than in any other text. New dimensions should also be examined in the evaluation and use of new information sources and types of features. Although traditional text is considered to be essentially static, it is quite dynamic and requires interaction with the numerous players that information offered in social media, such as online discussion forums, blogs, and Twitter posts. This may be considered not only as an additional source of complexity that can interfere with typical resuming methodologies but also as an opportunity to create an extra context that can help in the synthesis or make totally new forms of conceivable summary. For example, propose summarizing a blog post using information user reviews by extracting representative phrases. In [26], the author uses time correlation to extract relevant tweets to summarize events through a stream of tweets [27]. Addressing a summary by temporally extracting representative people, actions, and notions in Flickr data of the social network itself and not the substance of posts or messages.

Therefore, as we have indicated, existing social media NLP technologies are confronted by problems because of nonstandard spelling, noise, limited features, and errors. Therefore, some NLP solutions were suggested for improving Twitter news clustering efficiency, including standardization, terms expansion, enhanced selection of features, and noise reduction. Proper names and language switches would need quick and accurate recognition of names by entities and language detection techniques in a sentence. The focus of recent research is on language analysis in social media for social behavior and socially conscious systems.

The network infrastructure can deduce geolocation information. In [28], the author uses geolocation databases to associate locations with IP addresses. Several data sets can be used to map the geographical location between IP blocks. Usually, they are correct at country level, but at city level, they are much less reliable. In [29], it became apparent that for the following reasons, these databases were not very dependable. First of all, most of the database articles relate only a number of popular countries (such as the United States). This produces an uneven representation of the country across IP blocks in the database. Second, the entries do not always reflect the IP blocks' initial assignment. The author in [4] used a Naïve Bayes classification to attain a superior location accuracy based on multi-source IP mapping.

Another way to geolocating social network members can be based exclusively on friend lists ("you are where your friends lie") or followers' relationships.

In many circumstances, an individual's social network is sufficient to disclose their position [4, 5], first to provide a model for the distribution of distances among friends; then they used this distribution to find a user's place. They also engage more routinely and commonly employ this distribution with individuals near to themselves. The negative part of the strategy is that all users have the same distance distribution as friends and do not take into account population density in each area, indicating that the usage of population density leads to a more accurate location recognition. The location of the Twitter profile users was also analyzed as an additional information source. An analysis has been presented of how people use the placement area.

The sentiment analysis is a burgeoning field for the processing of natural languages with research, including the classification of documents to learning the polarity of words and sentences. Given the constraints of the character of tweets, the classification of Twitter messages is most akin to the sentence-level analysis. However, Twitter sentiment analysis is a completely different task because of the informal and specialized languages employed in tweets and the very nature of the microblogging realm. The question is, how successfully features and strategies are transferred into the microblogging realm on more well-formed material. Just in the last year, several publications examined the feeling and buzz of Twitter. Additional scientists have started to investigate the use of speaker elements, but the results remain unchanged. Microblogging features are common (e.g., emoticons) and are also popular, but the value of existing resource feelings based on non-microblogging data has not been researched. Researchers have also started to explore other approaches to obtain training data automatically. Many investigations rely on emoticons to define their data. Their experiments contain only the classification of feelings/no sentiment and not the classification of polarity in two directions, such as ours. Others use hashtags for training data. In this second grade, we are applying the following machine-learning method to produce the best results. Table 9.2 shows the literature survey summary.

Naive Bayes

This classification includes the Bayes classification, illustrated on a major classification issue: the categorization of text, the classification of a label categorizing text from several labels. In this classification, we introduce the naive classification for Bayes algorithms. We concentrate on a shared categorization of texts, an analysis of feelings, ex-sentiment analyses, positive or negative direction for a particular item expressed by a writer [5, 6].

The most rapid and easy classification is Naive Bayes. Many scientists say the best results have been produced in this classification. When we locate a label for a particular tweet, all labels having the feature are likely to be the

Table 9.2 Literature survey summary.

Author	Title	Method/tools	Application/result	Context
Yuliyanti, Djatna & Sukoco. (2017)	Sentiment Mining of Community Development Program Evaluation Based on Social Media	Lexicon-based and machine learning	Success level of the community development program	Twitter
Martin-Domingo, Martin, & Mandsberg. (2019)	Social media as a resource for sentiment analysis of Airport Service Quality	Machine learning	Analyse airport service quality	Twitter account
Mansonr. (2018)	Social Media Analysis of User's Responses to terrorism using sentiment analysis and text mining	Lexicon-based	Most user view IS1S as a threat and fear	Twitter
Saragih & Girsang, (2017)	Sentiment Analysis of Customer Engagement on Social Media in Transport Online	Lexicon-based	Evaluate the business performance of online transport.	1-accbook and Twitter comments
Hassan, Hussain, Husain, Sadiq, Lee. (2017)	Sentiment Analysis of Social Networking Sites (SNS) Data using Machine Learning Approach for the Measurement of Depression	Machine learning	Kind the depression level of a person	Twitter and newsgroup
Joyce & Deng. (2017)	Sentiment Analysis of Tweets for the 2016 US Presidential Election	Lexicon-based and machine learning	Calculate sentiment expressed and compare with polling data to see the correlation	Twitter
Ikoro, Harmina, Malik, & Batista-Navarro. (2018)	Analyzing Sentiments Expressed on Twitter by UK Energy Company Consumers	Lexicon-based	Analyze energy provider company and the sentiment that users show	Twitter

most likely, and the label is selected. The accuracy of unigrams is 79.67% lower. The precision increases with the detection of denial (81.66%) or higher n-grams (86.68%). We find that with both negative and higher grammes, the accuracy is somewhat lower than that (85.92%). We also observe that the accuracy of double classification is worse than that for one step.

Natural Language Processing

Natural language processing is an artificial intelligence field where computers are intelligent and beneficial in their analysis, understanding, and deriving meaning from human language. Using NLP, developers may organize and arrange knowledge to do tasks, such as automated resumption, translation, identification of entities, relationship extraction, feelings analysis.

Natural language processing takes into consideration the hierarchical structure of a language in addition to the traditional word processing operations, which see text as merely a sequence of symbols: a phrase is created in numerous wordings and phrases, including long-run concepts. Natural language processing systems have long fulfilled useful duties, such as grammatical correction, conversion of speech to text, and automatic translation between language, when evaluating language for its meaning. Natural language processing is used for text analysis, which makes it possible for machines to understand how people speak. This interaction between people and the computer enables applications in the real world, such as automated summary texts, feel analysis, the extraction of issues, identifying named things, tagging portions, extracting links, tightening, and so on.

For text mining, machine translation, and automated query answer, NLP is frequently employed. Natural language processing is recognized as an informatics challenge. Human language is seldom accurate or spoken simply. Knowing the language of man involves not only understanding words but also concepts and how they make logical sense. Although language is one of the easiest to master for the human mind, it is difficult for computers to understand NLP because of the ambiguity of the language.

9.2.1 Techniques in Sentiment Analysis

A recurring neural network [3] is one way to feeling analysis. The advantage is that it performs better than normal neural networks in structured variable-input data prediction. The input layer contains the word vector bag in the present network design at a time t. The input layer is connected with the concealed data history layer. The hidden layer has recurrent connections and also output layer connections. Neurons are also present in the hidden and output layers to store values at a time t. This recidivism enables the network to be deeper than a standard neural system [3]. For their sentimental analysis, the author in [3] suggested a semi-monitored dual recurrent neural network. It is similar to a standard recurring neural network in which a longer memory can be covered over time. It is unusual because the output layer additionally provides recursive connections to further analyze the feeling. The language is interpreted by a recursive neural network [12]. The primary target for language understanding is the seminal meaning of words [12]. The network is trained instead of words with semitone labels. The language is modeled with the output, when the input is adjusted to anticipate the word below. In [4], Twitter-specific methods of sentiment analysis are addressed. Electronic products domain has been the focus. The analysis of sentiment by Twitter is different from the standard sentiment analysis because the limit of 140 characters is more slang and misspelled words. Preprocessing before extracting features is therefore required. Two steps are taken to pre-processing. First, elements like hashtags and emoticons are taken from Twitter. Positive or negative results are offered on the basis that they are positive or bad. Positive or negative hashtags are also supplied. After Twitter has removed particular characteristics, a unigram technique is utilized, and the tweet text simply as a collection of words is represented. The feature vector uses positive and negative tweets. The Naive Bayes, a Support Vector Machine, and a Maximum Classifier include other texting classification approaches addressed in [4]. All features in the vector are considered independent by Naive Bayes classificator. A hyper plane is used to split tweets in classes by the Support Vector Machine. In this classification procedure, unlike the Naive Bayes Classifier, the Maval Entropy classification does not take on the relations among features; in the feature vector, the relations between the part of speech tags, emotive keywords, and denial can be used. Twitter sentiment was analyzed with emphasis on predicting election results [8]. It was stated that Twitter sentiment analysis uses n-grams and partial-speech tagging algorithms. Retweets generated by each party are considered as part of the prediction procedure. In this paper, the following were employed to predict elections with great precision [6]: Naive Bayes, Support Vector Machine, Max Entropy, and a rear-strength neural network. In [1], a neural network with events was used to evaluate the response of the Internet user to certain stimuli. Multiple artificial neural networks, one for each mood and many cultures, are made up of a mood engine. The ANNs are trained with terms that characterize the mood lexicon. Both blogs and social networks collect messages. In [7], a recursive neural tensor network was constructed, which seeks to better understand a group of words' semantics. It was designed to understand short messages usually found on Twitter.

The recurrent neural tensor Network demonstrated the greatest precision when compared with standardized recurrent neural networks, matrix-vector recursive neural networks, Naive Bayes, and vector support machines [31].

9.3 Implementation and Results

The users first give input, i.e., a clave word for extraction of tweets and then categorize the extraction tweets by means of the idea that the traditional method for surveys can overcome the disadvantages of getting people's opinions, by training the model to classify tweets based on tweet feelings and to make the model as accurate as possible.

Get the Forecast of Tweet Feeling With Our Machine Learning Model

Machine learning model predicts the tweets' feeling at this stage, but the results are saved in an array.

Show the graphic representation of results

This stage will provide the end users the results in a graphical way, for a better comprehension of the results, such as a bar graph or diagram. A diagram that shows the working flow of the stepwise activities and the action with support for choice, iteration, and competition is another representation of the software. It shows a set of operations or control flows in a flowchart or data flow diagram related system. The process of building and training the machine learning model for prediction is shown in Figure 9.3.

9.3.1 Online Commerce

Electronic commerce is the most common usage of feelings analysis. Websites enable people to provide their shopping and product quality experience. They summarize the product and its various aspects by providing ratings or scores. It is easy for customers to view thoughts and recommendations on the entire product and its characteristics. The users will be given with a graphical summary of the total product. Popular commercial websites, such as amazon.com feature reviews from publishers and customers. http://tripadvisor.in, is a prominent hotel and tourist destination review site. It has 75 million reviews and opinions around the world. Feeling analysis aids such websites through the study of the vast volume of views by transforming unmet customers into boosters.

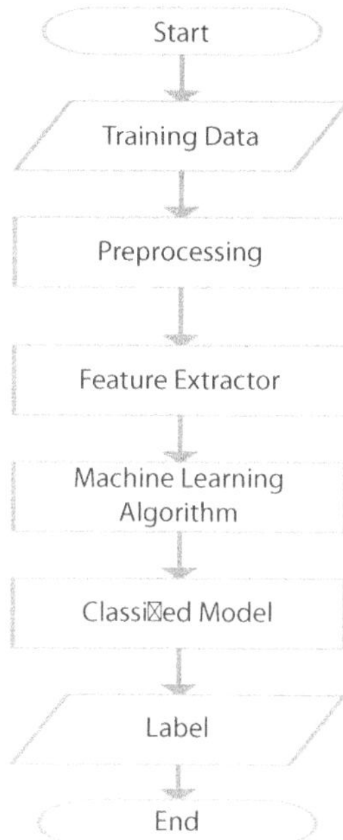

Figure 9.3 Workflow of proposed system model.

9.3.2 Feature Extraction

The features discovered in the tweet are listed in the following paragraphs to extract the feeling and how they are being treated.

9.3.3 Hashtags

These capabilities allow you to organize tweets in a thread tied to a given topic. Inclusion of a hashtag, like #android, might express good or negative mood preferences, which would allow the functionality to be employed. Each field has the best ranking hashtags, and their existence adds weight to the feature, leaning it to one of the classes.

9.3.4 Punctuations

Exclamation points, inverted commas, question marks, and unclear punctuation are all weighted to indicate a characteristic. This characteristic is a weighted average of all punctuation contributions in a review.

9.4 Conclusion

The task of sentimental analysis is still in the development stage and far from complete, particularly in the field of microblogging. We, therefore, want to put forward a few ideas that we feel should be explored and can lead to a further improvement in performance. We have worked just on extremely simple unigram models for now. By integrating further information, such as word proximity with a word denial, we could improve those models. The window could be indicated before to the word (e.g., a 2- or 3-word window), and if it lies within the windows, the negative effect may be included to the model.

The closer the word denial is to the word unigram, which has to be computed with its previous polarity, the more polarity it should affect. If the negation lies close to the word, for example, the polarity of the word can simply be reversed, the further the negation is from the word, the more minimal if the effect. Furthermore, we focus primarily on the unigrams and investigations of the effects of bigrams and trigrams. As indicated by bigrams, this typically improves performance, as is stated in the study review section. However, we require a far larger data collection than our 9,000 tweets, so that we can make bigrams and trigrams a successful feature. Instead of estimating one single probability, we might have several probabilities for each word as P(word | obj), depending on the part of the speech to which that word is associated. Using a quite similar approach, it asserts that the POS data appended to each unigram does not result in any substantial performance difference (with Naive Bayes having somewhat greater efficiency, while SVM's performance decreases small). However, these results have been confirmed to classify reviews and to analyze the feelings of services, such as Twitter. Another issue that we have to examine is whether the information concerning the relation of the term in a tweet affects the categorization system performance, examines a similar characteristic, and reports negativity. We concentrate on generic sentiment analysis in this research. We discovered, for example, that consumers usually utilize specific types of keywords that may be classified in a few classes, namely, media/films/music, celebrities, products/brands, sportsmen, politicians. So, we may try to conduct distinct tweet feelings that pertain to one of the courses exclusively and compare the findings if we do an overall feeling analysis on the results. The training data would not be generic but would be specialized for one of those classes (Figures 9.4–9.7).

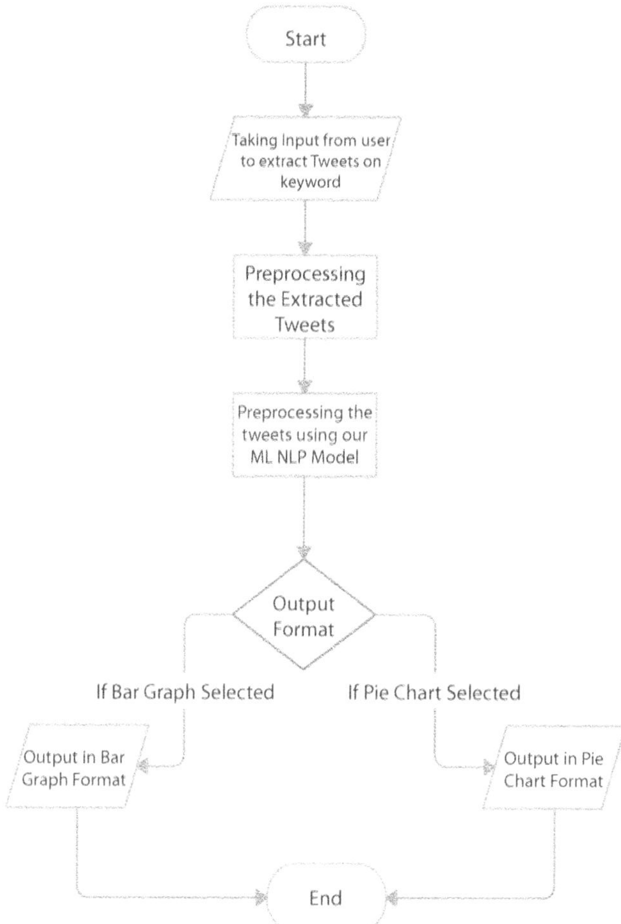

Figure 9.4 The user-side program's activity diagram.

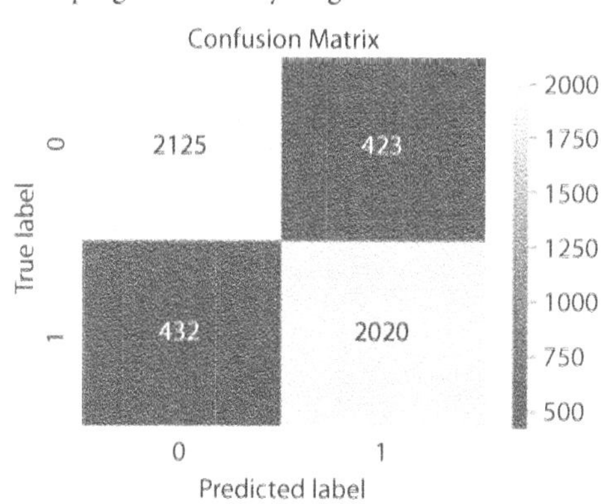

Figure 9.5 Confusion matrix with 92.50% accuracy.

Figure 9.6 Hashtag/keyword entry page.

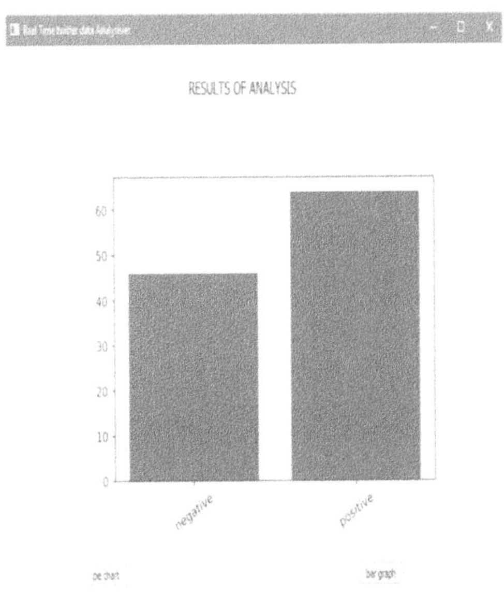

Figure 9.7 An examination of the effectiveness of positive and negative tweets.

9.5 Future Scope

Currently, the Naïve Bayes algorithm is used for this work, and it is one of the machine learning algorithms, which has an exactness of just about 92.50%. To strengthen the preciseness of our model and to better forecast the model, we will be examining and applying the deep learning algorithms to our NLP model in future.

References

1. Shitharth, S., Manimala, K., Bhavani, V.V., Nalluri, S., Real time analysis of air pollution level in metropolitan cities by adopting cloud computing based pollution control monitoring system using Nano Sensors. *Solid State Technol.*, 63, 2, 1031–1045, 2020.
2. Sangeetha, K., Venkatesan, S., Shitharth, S., Security Appraisal conducted on real time SCADA dataset using cyber analytic tools. *Solid State Technol.*, 63, 1, 1479–1491, 2020.
3. Thirumaleshwari Devi, B. and Shitharth, S., An Appraisal over Intrusion Detection systems in cloud computing security attacks, in: *2nd International Conference on Innovative Mechanisms for Industry applications*, IEEE Explore, p. 122, 2020.
4. Sangeetha, K., Venkatesan, S., Shitharth, S., A Novel method to detect adversaries using MSOM algorithm's longitudinal conjecture model in SCADA network. *Solid State Technol.*, 63, 2, 6594–6603, 2020.
5. Shitharth, S., Satheesh, N., Praveen Kumar, B., Sangeetha, K., IDS Detection Based on Optimization Based on WI-CS and GNN Algorithm in SCADA Network, in: *Architectural Wireless Networks Solutions and Security Issues*, Lecture notes in network and systems, vol. 196, 1, pp. 247–266, 2021.
6. Mohammad, G.B., Shitharth, S., Kumar, P.R., Integrated Machine Learning Model for an URL Phishing Detection. *Int. J. Grid Distrib. Comput.*, 14, 1, 513–529, 2021.
7. Mohammad, G.B. and Kandukuri, P., Detection of Position Falsification Attack in VANETs using ACO. *J. Int. J. Control Autom.*, 12, 6, 715–724, 2019.
8. Madhuri, T. and Mohammad, G.B., A Supervised Learning Based Recommender System for Breast Cancer Prognosis in India. *Int. J. Control Autom.*, 12, 6, 753–767, 2019.
9. Palaniappan, S. and Awang, R., Intelligent Heart Disease Prediction System Using Data Mining Techniques. *IJCSNS Int. J. Comput. Sci. Netw. Secur.*, 8, 8, 343–350, 2020.

10. Patil, B. and Kumaraswamy, Y.S., Extraction of Significant Patterns from Heart Disease Warehouses for Heart Attack Prediction. *IJCSNS Int. J. Comput. Sci. Netw. Secur.*, 9, 2, 228–235, 2019.
11. André, Bernstein, M., Luther, K., Who gives a tweet?: Evaluating microblog content value, in: *Proc. of the ACM Conference on Computer Supported Cooperative Work (CSCW)*, pp. 471–474, 2012.
12. Artstein, R. and Poesio, M., Inter-coder agreement for computational linguistics. *Comput. Linguist.*, 34, 553–596, 2008.
13. Venugopalan, M. and Gupta, D., Exploring Sentiment Analysis on Twitter Data. *Eighth International Conference on Contemporary Computing (IC3)*, pp. 1–7, 2015.
14. Duncan, B. and Zhang, Y., Neural Networks for Sentiment Analysis on Twitter, in: *14th International Conf. on Cognitive Informatics & Cognitive Computing*, pp. 275–278, 2015.
15. Neela Anupama, B.S., Rakshith, D.B., Rahul Kumar, M., Navaneeth, M., Real Time Twitter Sentiment Analysis using Natural Language Processing. *Int. J. Eng. Res. Technol. (IJERT)*, 9, 7, 1107–1112, 2020.
16. Virmani, C., Pillai, A., Juneja, D., Extracting Information from Social Network using NLP. *Int. J. Comput. Intell. Res.*, 13, 4, 621–630, 2017.
17. Backstrom, L., Sun, E., Marlow, C., Find me if you can: Improving geographical prediction with social and spatial proximity, in: *Proc. of the 19th International Conference on World Wide Web*, pp. 61–70, 2010.
18. Farzindar, A. and Inkpen, D., *Natural Language Processing for Social Media*, pp. 41–82, Morgan & Claypool Publishers series, San Rafael, CA, USA, 2017.
19. Balahur, A., Sentiment Analysis in Social Media Texts. *Proceedings of the 4th Workshop on Computational Approaches to Subjectivity, Sentiment and Social Media Analysis*, pp. 120–128, 2013.
20. Baldwin, T., Cook, P., Lui, M., MacKinlay, A., Wang, L., How noisy social media text, how different social media sources, in: *Proc. of the 6th International Joint Conference on Natural Language Processing*, pp. 356–364, 2013.
21. Baldwin, T. and Li, Y., An in-depth analysis of the effect of text normalization in social media, in: *Proc. of the Conference of the North American Chapter of the Association for Computational Linguistics: Human Language Technologies*, pp. 420–429, 2015.
22. Balikas, G. and Amini, M.-R., TwiSE SemEval-2016 task 4: Twitter sentiment classification, in: *Proc. of the 10th International Workshop on Semantic Evaluation (SemEval-2016)*, pp. 85–91, 2016.
23. Barbier, G., Feng, Z., Gundecha, P., Liu, H., Provenance data in social media, in: *Synthesis Lectures on Data Mining and Knowledge Discovery*, pp. 1–55, Morgan & Claypool Publishers, San Rafael, CA, USA, 2013.
24. Barbieri, F., Saggion, H., Ronzano, F., Modelling sarcasm in Twitter, a novel approach, in: *Proc. of the 5th Workshop on Computational Approaches to Subjectivity, Sentiment and Social Media Analysis*, pp. 50–58, 2014.

25. Barman, U., Das, A., Wagner, J., Foster, J., Code mixing: A challenge for language identification in the language of social media, in: *Proc. of the 1st Workshop on Computational Approaches to Code Switching*, pp. 13–23, 2014.
26. Baroni, M., Chantree, F., Kilgarriff, A., Sharoff. Cleaneval, S., A competition for cleaning web pages, in: *Proc. of the 6th International Conference on Language Resources and Evaluation (LREC'08)*, pp. 1–8, 2011.
27. Nithyashree, T. and Nirmala, M.B., Analysis of the Data from the Twitter account using Machine Learning. *5th International Conference on Communication and Electronics Systems (ICCES)*, pp. 989–993, 2020.
28. Meena, R. and Thulasi Bai, V., Study on Machine learning based social media and Sentiment analysis for medical data applications. *Third International conference on I-SMAC (IoT in Social, Mobile, Analytics and Cloud) (I-SMAC)*, pp. 603–607, 2019.
29. Hu, T., She, B., Duan, L., Yue, H., Clunis, J., A Systematic Spatial and Temporal Sentiment Analysis on Geo-Tweets. *IEEE Access*, 8, 8658–8667, 2020.
30. Mahamood, Md. R. and Shilpa, B., Design and Implementation of Framework for Smart City using Lora Technology. *Sreyas Int. J. Sci. Technocr.*, 1, 11, 36–43, 2017.
31. Gouse, G.M. and Ahmed, A.N., Ensuring the public cloud security on scalability feature. *J. Adv. Res. Dyn. Control Syst.*, 11, 1, 132–137, 2019.

10

Cascading Behavior: Concept and Models

Bithika Bishesh

School of Business Studies, Sharda University, Greater Noida, India

Abstract

Cascade networks are an example of networks that connect individuals on the basis of the direction in which the data or information flows between them. These networks have garnered the attention of various sociologists interested in the diffusion of innovation for many years. Research goals have shifted over time and across platforms, from simply seeing and counting cascades to tracking, anticipating flow of information, and modeling them. Thus, understanding cascades is a crucial step in gaining a better understanding of how information spreads. This chapter will give an overview of the cascading behavior. It begins with an introduction to networks and the graph theory then move on to discuss cascade networks in depth along with their purpose and significance. The concepts of centrality, cascading failure, and cascading capacity are also covered. The different models that are discussed are decision-based models, probabilistic models, independent cascade model, linear threshold model, and Susceptible, Infectious, or Recovered (SIR) model. In the end, cascading behavior is explained using python codes to illustrate the practical applications of this concept. The focus is on the application of this concept to various substantive examples.

***Keywords*:** Cascading behavior, networks, graph theory, centrality

10.1 Introduction

In their broadest meaning, networks are structures made up of a collection of nodes and links. The connections connect the nodes, embodying a certain sort of relationship between them. A social network consists of people as nodes and links demonstrating the relationship among these people [1].

Email: bithika.bishesh@sharda.ac.in

Mohammad Gouse Galety, Chiai Al Atroshi, Bunil Kumar Balabantaray and Sachi Nandan Mohanty (eds.) Social Network Analysis: Theory and Applications, (175–204) © 2022 Scrivener Publishing LLC

This focus on conceptual outlook of human relationships makes the analysis of social networks intriguing by giving an insight into different types of social relationships.

There are several examples of complicated systems everywhere around us. Social networking sites, transport systems, disease transmission are all forms of complex systems that we encounter everyday. Because they comprise of multiple apparently separate elements that might be directly or indirectly related, all of these systems are very difficult to comprehend. These components or elements display difficult to understand behavior and are far too dangerous to experiment with. Even a minor, relatively harmless alteration in one of the elements can set off a cascade of events. Acquaintance, coauthors, colleagues, affiliation, personal relationships, friends, information sharing, and so on are some examples of such interactions. All of these networks connect individuals through a link by which they communicate with one another for reasons, such as information exchange and cooperating.

The key notion of networks is their "connectedness." This is found in domains, such as biology, social studies, and, computer science. Networks are represented mathematically as graphs containing nodes and edges or link.

The history of graph theory traces its origin to a famous story from the 18th century. A river flowing through a town in Russia created four different islands, which are shown in Figure 10.1. In Figure 10.1, we can see that there are four different islands A, B, C, and D, and these four islands are connected by seven bridges. Islands A and B are connected by bridge numbers 1 and 2, islands A and C are connected by bridge numbers 3 and 4, islands A and D are connected by bridge number 5, island C is connected to island D with bridge number 6, and island D is connected to island B with bridge number 7. A question was raised whether it is possible to visit

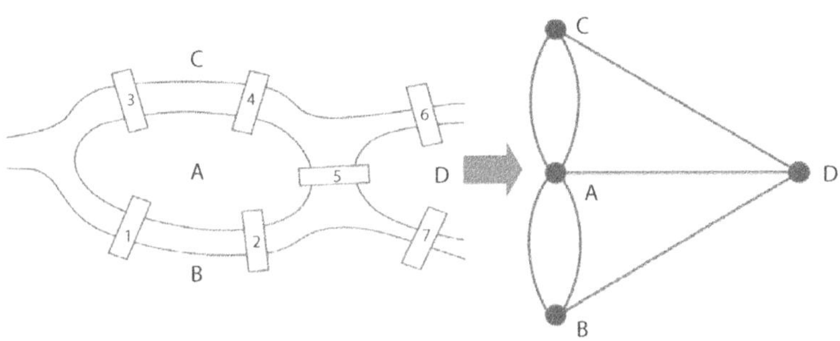

Figure 10.1 Concept of graph theory.

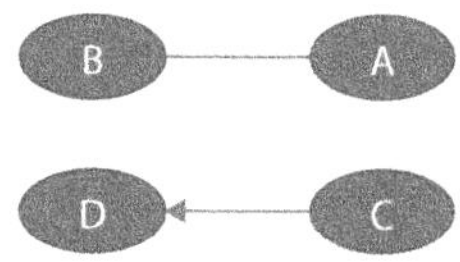

Figure 10.2 Connections in a graph.

each of these islands by crossing the bridge only once. Euler came with a solution to this problem in 1735 stating that it is impossible because each bridge is to be crossed only once, therefore, the sequence in which this is done is very important. This led to the birth of the graph theory. As per this theory, all the islands are considered as a vertex, and all the different bridges are considered as edge.

Figure 10.1 shows the image for graph theory wherein the points A, B, C, and D are the vertices or the nodes, and they are connected with each other through one or more nodes.

The basic foundation of a graph is that it consists of related object pairs, which correspond to vertex or nodes, and each vertex is connected to each other through an edge or link [2]. There are two different ways in which this connection happens. It could be either bidirectional or directional as shown in Figure 10.2. If every edge is directional, then it is called a directional graph. The two nodes, C and D, show a bidirectional relationship, wherein C has some relation or link with D but the opposite is not true, unless explicitly mentioned. Nodes A and B are part of an undirected graph because they have a bidirectional line.

10.2 Cascade Networks

Cascade networks are an example of such networks that connect individuals on the basis of the direction in which the data or information flows between them [3]. As per economists, information cascade is when a person observes the behavior of other individuals and considers it best to follow their behavior, without taking his own information into consideration. Essentially, a clear distinction is made by the economists between herding behavior and information cascade. This distinction lies in the fact that an information cascade is where people make decisions based on inferences while disregarding their own data, whereas in herding, they follow the "herd" without particularly overlooking their own data. Messages spread in cascades from one person to another via social network links. The path taken by these messages in travelling from one person to another accumulate and generate a network is called cascade networks. These networks lie on top of the social network as a layer, and the paths are commonly referred to as information paths.

A cascade consists of a seed individual communicating information independently of others, accompanied by others who are inspired by the seed in sharing the same information. These are periods in which members of a group display herd-like behavior as a result of their judgments, depending on the actions of others and not on their own knowledge. Cascades are frequently viewed as unique expression of many strong but complex and weak systems because a system may look steady for a considerable amount of time and tolerate several external shocks, only to suddenly produce a massive cascade.

Information cascades are when a behavior or concept spreads rapidly as a result of the influence of others. The term "cascade" is also used to describe "fads" or "resonance." These networks have garnered the attention of various sociologists interested in the diffusion of innovation for many years; more recently, researchers in a variety of fields have looked into cascades for the purposes of identifying trendsetters for viral marketing, identifying inoculation targets in epidemiology, and also explaining trends in blog space. Despite extensive empirical work in sociology on data sets of modest size, data availability has reduced the scope of analysis.

10.3 Importance of Cascades

Understanding cascades is a crucial step in gaining a better understanding of how information spreads on the Internet. This spread of information gives us vital information about the people who are participating. Cascades, as previously said, reflect some sort of link between the users. Influence is a term used in the literature to describe this relationship [5]. Identifying influencers has gotten a lot of attention in the past, and cascades have been used as markers of influence. As a result, many studies recognize the paths that information travels to reach individuals as influence paths, because they directly suggest that one user encouraged or influenced another in propagating the message.

Apart from influence, researchers have discovered that sharing the same content is motivated by a sense of homophily among the participants. Repeated exposure to a piece of content raises the likelihood of it being shared. They reasoned that these users are vulnerable to both influence and homophily in this situation; continued exposure raises the influence factor, and being around a group of users who are prone to an item suggests that the user is susceptible as well. Cascades, on the other hand, do not arise just as a result of influence and homophily, as both are linked to the nature of content.

Thus, a cascade tells us about the content's value. Considering that users have narrow attention span, the most successful cascade is one that attracts the most attention among competing cascades at any given time. Because they are built utilizing a portion of the social network, cascade networks are referred to as implicit networks. Connections on social networks indicate that users are eager to listen to one another; however, cascade networks have much better indicators because they are formed by a forced sharing action that propels the content to the friend list of the users.

Users frequently build new links because of their exposure to new sources of information and such an analysis of cascades can also help in identifying link creation and evolution of networks.

10.4 Purposes for Studying Cascades

Research goals have shifted over time and across platforms, from simply seeing and counting cascades to tracking, anticipating flow of information, and modeling them. The most important goal is to track existing cascades before either constructing or inferring new ones. The capacity to build a cascade is totally dependent on the data available during the process. The second viewpoint focuses on architecturally, chronologically, or simply quantitatively measuring cascades, in combination with various platform-dependent measurements.

Very often, tracking cascaded is the first step before measuring them. The structural study of cascades, for example, necessitates the construction of cascades prior to the analysis phase. The third viewpoint examines cascade modeling or the use of generative algorithms to generate cascade networks based on the properties found in monitored cascade networks. The fourth perspective looks into things like whether or not a piece of information will be shared, whether or not a popular piece of content will remain popular, and how to forecast the growth of a cascade.

10.5 Collective Action

Individuals who are self-interested logically pursue their goals by taking into account the several limits imposed by their external environment and when extended to collective action, theories of economics predict undersupply.

Conversely, mass behavior is defined as exuberant, emotional, illogical, suggestible, hypnotic, chaotic, and unpredictable so here collective action appears to be oversupplied. Collective action's infectious and chaotic dynamics seem to defy rationalization.

Most studies of collective action agree that addressing the free rider dilemma necessitates grouping potential contributors and intertwining their decisions. The ability to organize is determined by the group's social links, namely their general density or frequency of ties, the degree to which they are centralized in a handful of people, and the expense of communicating and coordinating actions via these ties. The study of collective behavior, such as market dynamics, the establishment of social norms and conventions, and collective phenomena in everyday life, such as traffic congestion, is gaining popularity.

10.6 Cascade Capacity

With reference to an endless network, the capacity of a cascade network is described as the highest threshold, which triggers a full cascade by a set of finite nodes [6]. Suppose there is a network that has unlimited number of nodes, each of which is linked to a fixed number of others. Further, suppose a small number S of such nodes exhibit behavior A, whereas the remaining displays B, all such nodes that were showing behavior B in the beginning now adopt behavior A on the basis of the threshold in each time step t. As a result, the network's cascade capacity is defined as the highest value of q at which S will produce full cascade.

10.7 Models of Network Cascades

Networks can be thought of as a conduit via which information or contagion travels. It can be viewed from a variety of perspectives—cascading behavior, network effects, diffusion of innovations, or failures. The current pandemic has made the study of this behavior more vital. Moreover, it also helps in getting an understanding of how crucial things, like fake news, viral marketing, are in the digital age.

Diffusion models are divided into two categories—decision-based models and models that are probabilistic.

We give nodes the ability to make decisions in decision-based models. A node examines the behavior of its neighbors before making its own decision. This includes things like the adoption of new technologies or protest rallies. In probabilistic models, infected node strives to spread the infection to an uninfected node which is particularly important to understand in the event of a pandemic.

10.7.1 Decision-Based Diffusion Models

Let us take a game theoretic approach to this. The game is based on a two-player cooperation game. A player can adopt either behavior A or B [7]. The essential insight is that if your peers have adopted the same behavior as you, you will receive more payoff. Each participant is given the freedom to play his or her own game and maximize his or her payoff.

Consider the payoff matrix below for two node behaviors:

- Both the nodes will get a payoff a greater than zero if they adopt behavior A,
- Both the nodes will get a payoff b greater than zero if they adopt behavior B and,
- Both the nodes will get a payoff zero if they adopt the opposite behaviors.

When the network is large, every node is mimicking this game with every neighbor and the final payoff is the total of payoff of all the nodes over all the games. Suppose a node in a very large network has d neighbors. If p is the fraction of neighbors of v adopting decision A and rest decide on decision B, then v's total payoff is:

$$\text{payoff}_v = \begin{cases} a.p.d & \text{if v chooses A} \\ b.(1-p).d & \text{if v chooses B} \end{cases}$$

Thus, v selects A if $p > \frac{b}{a+b} = q$, where p is the proportion of v's neighbors, and A and q are the payoff threshold.

10.7.2 Probabilistic Model of Cascade

Such types of cascades are most commonly associated with epidemics where contagious diseases spread from one individual to another. The diseases can transmit rapidly across the population or remain undetected for a long time. The cascade of illness is analogous to cascade behavior in certain ways. For instance, they have a common feature that they pass from one individual to the next. The similarities, however, stop there. Adoption of a behavior is determined by a person's decision to switch to that behavior. In the case of disease propagation, however, a person's decision is irrelevant. In epidemics, only a person's vulnerability is a factor.

A contact network can be used to show the transfer of diseases from one person to another. In such type of network, each node represents an individual, with an edge in between individuals when they get in contact, allowing illness to spread. These networks are also used to trace the propagation of dangerous software over communication networks, as well as to investigate the transmission of fast-moving illnesses in plants and animals [8].

The pathogen and the network are inextricably linked: various diseases afflicting the same population may spread at different rates, depending on the transmission medium. Airborne illnesses would propagate quicker and further, infecting a substantial portion of the population in a short period. When compared with other means of transmission, a disease that requires some type of physical touch may not spread as quickly or infect as many people.

Because there is a certain probability connected with the dissemination of a disease at each step of contact, the cascade process in the case of epidemics is much more arbitrary than the one employed in that of the modeling behavior. There are numerous models that can be used to explain such arbitrary behavior, which are addressed below.

Independent Cascade Model and Linear Threshold Model—We can differentiate between active, or infected, nodes, known as seeds, that propagate information, and inactive nodes in both models. Recursively, the already infected nodes have a chance of infecting their neighbors. After a given number of these kind of cascading cycles, the network becomes infected with a large number of nodes. The principle behind LTM is that a node becomes active when a large portion of its neighbors also become active. In more formal terms, the threshold for every node u is t, that is determined at random from the range [0, 1]. The threshold is the proportion of u's neighbors, which must activate before u can become active.

A very small number of nodes in a particular network are set to active at the start of the procedure to allow the information diffusion process to begin. If the fraction of active neighbors is greater than the threshold, a node activates in further steps of the process. Instead, in Independent Cascade Model, a seed u attempts to influence one of its inactive neighbors; however, node u's success in activating node v is solely dependent on the edge propagation probability from u to v (since every edge has its own given value). Irrespective of whether or not it succeeds, the same node will never have another opportunity to activate the very same inactive neighbor. When no more nodes are activated, the procedure ends.

10.7.3 Linear Threshold Model

We have the given setup in a Linear Threshold Model:

- A node v having random threshold which is = $\theta v \sim U[0, 1]$
- Node v which is getting influenced by every neighbor w as per the weight bv,w, so that

$$\sum_{w \text{ neighbour of } v} b_{v,w} \leq 1 \tag{10.1}$$

- A node v gets active if at least θv fraction of this node's neighbors are active i.e.:

$$\sum_{w \text{ active neighbour of } v} b_{v,w} \leq \theta_v \tag{10.2}$$

Figure 10.3 shows this process where node V becomes active, influencing U and W by 0.2 and 0.5, respectively; node W becomes active, influencing U and X by 0.3 and 0.5, respectively; node U becomes active, influencing X & Y by 0.1 & 0.2; (D) node X becomes active, influencing Y by 0.2; the process ends when it is impossible for any other node to activate [9].

10.7.4 Independent Cascade Model

In this type of model, the influences (activation) of nodes are modeled on the basis of the probabilities in a directed graph as shown in Figure 10.4:

- A directed graph which is G=(V,E)
- Given S is a node set that begins with a new behavior and they are active
- Every edge (v,w) has a probability of p_{vw} and fires only once
- If node v activates, it has only one chance of activating currently inactive neighbor w, and the probability of doing this is p_{vw}
- Activation spread through the network
- If both u and v are active and linked to w, it makes no difference which of them attempts to activate w first.
- This deterministic model is unique case of the Independent Cascade model. Here, for all (v,w), p_{vw} =1

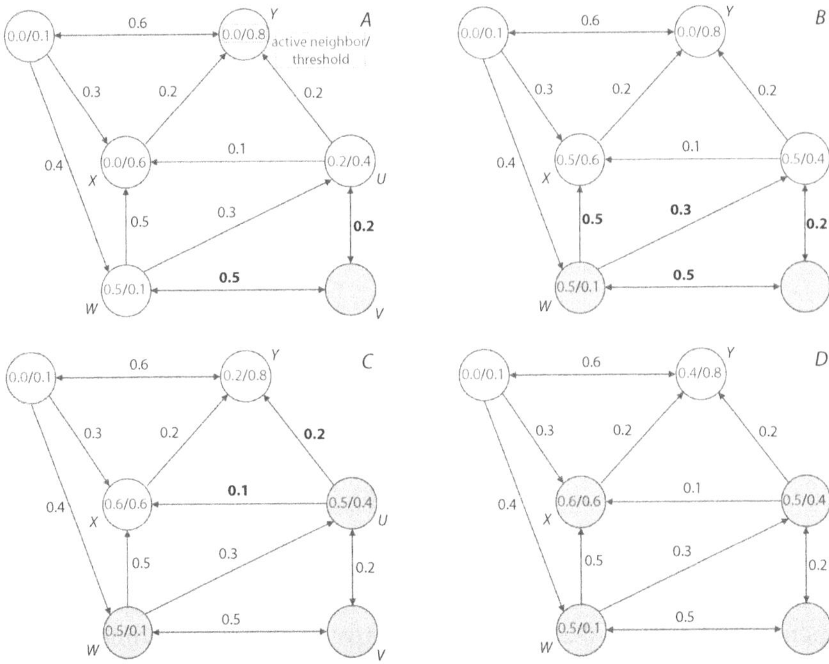

Figure 10.3 Linear threshold model. Source: https://snap-stanford.github.io/cs224w-notes/network-methods/influence-maximization.

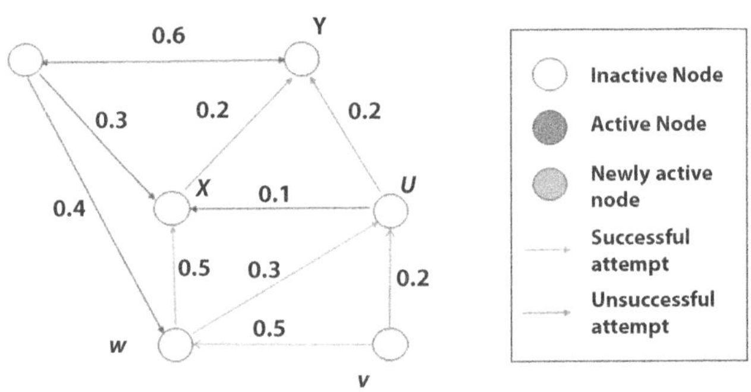

Figure 10.4 Independent cascade model.

10.7.5 SIR Epidemic Model

The following is how the model unfolds: At first, only a small amount of nodes are infected, whereas the rest are susceptible. For a set number of steps tI, every node which is infectious stays the same. Following tI time, v node enters state "Removed" and, is now unable to infect or get infected [10].

Susceptible (S_t): The individual is prone to infection from his neighbors but not yet infected.

- Infected (I_t): Infected individuals having a chance of infecting their other neighbors who are susceptible.
- Removed (R_t): Once the individual has completed the entire infectious time, he is no longer kept in the network because they have become immune. This model is the basic mathematical notation for the transmission of disease which divides the entire population consisting of N individuals into three major compartments and these differ as a function of time t.

The SIR model uses two parameters, β and γ, to characterise the alteration in the population of each of such compartments. β describes the disease's effective contact rate. S/N is the proportion of individuals prone to getting infected. Per unit period, an infected person comes into contact with βN additional people. The average recovery rate is γ, and the average time it takes for an infected person to transfer the illness on is 1/γ.

The differential equation for this model is represented as Kermack and McKendrick [4]:

$$\frac{dS}{dt} = -\frac{\beta SI}{N}$$
$$\frac{dI}{dt} = \frac{\beta SI}{N} - \gamma I \qquad (10.3)$$
$$\frac{dR}{dt} = \gamma I$$

To understand this SIR model with the help of python code as shown in Figure 10.5, let us suppose that for the spread of a disease in a town with ten thousand individuals, the effective contract rate is 0.4, and the average time it takes for an infected person to transfer this illness on is 14 days.

Thus, the parameters are:

β=0.4,
1/γ=14
N=10,000
The model begins with one infected individual on day 0 and so:
I(0)=1.

Figure 10.6 depicts the SIR model graph using python.

```
import numpy as np
from scipy.integrate import odeint
import matplotlib.pyplot as plt

# Total population, N.
N = 10000
# Initial number of infected and recovered individuals, I0 and R0.
I0, R0 = 1, 0
# Everyone else, S0, is susceptible to infection initially.
S0 = N - I0 - R0
# Contact rate, beta, and mean recovery rate, gamma, (in 1/days).
beta, gamma = 0.4, 1./14
# A grid of time points (in days)
t = np.linspace(0, 160, 160)

# The SIR model differential equations.
def deriv(y, t, N, beta, gamma):
    S, I, R = y
    dSdt = -beta * S * I / N
    dIdt = beta * S * I / N - gamma * I
```

Figure 10.5 SIR model in python. *(Continued)*

10.8 Centrality

The main measures of centrality are categorized as:

- radial—the number of paths that originate from a particular node,
- medial—number of some paths that pass through a particular node.

Degree Centrality—Degree is the simplest of all the measures to calculate. It is the total number of incident edges for a particular node. Because degree shows the number of path that originates from a node with length one, it is categorized as a radial measure [11].

Shell Number—Shell number, also known as "k-shell number," is another type of radial measure that is calculated by shell decomposition. In the network of an SIR model, large shell-number nodes are called as the "core"

```
    dRdt = gamma * I
    return dSdt, dIdt, dRdt

# Initial conditions vector
y0 = S0, I0, R0
# Integrate the SIR equations over the time grid, t
ret = odeint(deriv, y0, t, args=(N, beta, gamma))
S, I, R = ret.T

# Plot the data on three separate curves for S(t), I(t) and R(t)
fig = plt.figure(facecolor='w')
ax = fig.add_subplot(111, facecolor='#dddddd', axisbelow=True)
ax.plot(t, S/10000, 'b', alpha=0.5, lw=2, label='Susceptible')
ax.plot(t, I/10000, 'r', alpha=0.5, lw=2, label='Infected')
ax.plot(t, R/10000, 'g', alpha=0.5, lw=2, label='Recovered with immunity')
ax.set_xlabel('Time /days')
ax.set_ylabel('Number (1000s)')
ax.set_ylim(0,1.2)
ax.yaxis.set_tick_params(length=0)
ax.xaxis.set_tick_params(length=0)
ax.grid(b=True, which='major', c='w', lw=2, ls='-')
legend = ax.legend()
legend.get_frame().set_alpha(0.5)
for spine in ('top', 'right', 'bottom', 'left'):
    ax.spines[spine].set_visible(False)
plt.show()
```

Figure 10.5 (Continued) SIR model in python.

and are viewed as influential spreaders. This technique has also been used in the real world to locate crucial nodes in a network. Shell decomposition can be used to discover crucial nodes in a subgroup of independent systems on Internet, or to detect persons who are likely to launch information cascades on online social network.

Betweenness Centrality—High betweenness centrality nodes act as "bottlenecks" because several paths in the network travel via them. As a result, betweenness is a measure of medial centrality.

Suppose for two nodes s and t, the number of shortest paths is *σst* and the number of shortest paths among these two nodes containing *v* node is *σst(v)*. Then for node v, the betweenness centrality is given by:

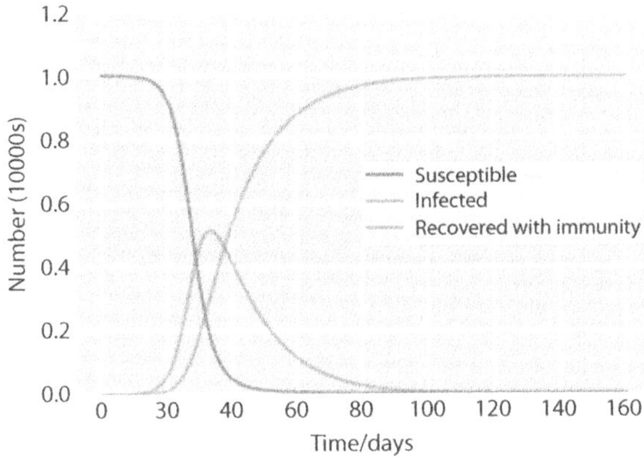

Figure 10.6 SIR model graph showing susceptible, infected & recovered individuals.

$$\sum s\_{\neq}v{\neq}t\frac{\sigma st(v)}{\sigma st} \tag{10.4}$$

Closeness Centrality—It is the inverse of the mean shortest length of the path from a node i to all of the other nodes in a given network. Closeness, on the surface, refers to closeness of a node to other nodes. For two nodes i and j, if the shortest path is given as function dG(i, j), then the mean length of path from node i to all the other nodes is:

$$\boldsymbol{L_i} = \frac{\sum_{j\epsilon V\backslash i} \boldsymbol{d_G}(i,j)}{|V|-1} \tag{10.5}$$

So, closeness of a node can be represented as:

$$\boldsymbol{C_c}(\boldsymbol{i}) = \frac{1}{\boldsymbol{L_i}} = \frac{|V|-1}{\sum_{j\epsilon V\backslash i} \boldsymbol{d_G}(i,j)} \tag{10.6}$$

Eigenvector Centrality—For network V = (G,E) having surrounding matrix A = (aij), wherein aij is equal to one if there exists an edge among the two nodes i and j, the eigenvector centrality is:

$$\boldsymbol{x_i} = \frac{1}{\lambda}\sum_{j\in V} \boldsymbol{a_{ij}x_j} \tag{10.7}$$

for some λ. Now, if for x_i, its vector is considered to be x then the relationship in equation can be written as:

$$x = \frac{1}{\lambda} Ax, \text{or } Ax = \lambda x \tag{10.8}$$

This equation relates *A* with its eigenvector and eigenvalues.

10.9 Cascading Failures

Cascading failure is a type of failure in a system with interconnected parts wherein if one part fails, it will lead to failure of other subsequent parts as well. Here, the normal working state is θ=N, and the final failure state is state θ =N. Computer networks and electrical systems are both susceptible to this type of failure. The intermediate states *i*, $2 \leq i \leq N-1$ are compromised states wherein if one system fails, it will lead to failure of other systems. Failure and compromised states are assumed to be irreversible, meaning that the system cannot be repaired or restored to its original condition soon after a defect occurs.

10.10 Cascading Behavior Example Using Python

Now let us look at the problem of information cascade, which is like a set of initial adopters of some decision in the network, and we have to look at how soon the decision will travel in that network. For instance, in the example given in Figure 10.7, *v* and *w* are the initial adoptors for an individual. We have to find out whether v and w are significant in influencing the behavior of other individuals in the same network.

This network-wide coordination game has two evident equilibria in each network: one where everybody adopts A, and another where everybody adopts B. Every node is constantly revising its decision based on what its neighbors are doing right now. Assume that B is the default behavior for everyone in the network at first. Then a small group of "initial adopters" also decide to use A.

Suppose that the initial adopters moved to A for some inexplicable reasons. Because the initial adopters have switched to A, a few of their neighbors may also do the same, and so on, in possibly a cascading pattern.

The procedure ends either when each node switches to A or when no node wishes to switch, at which time the situation has stabilized on A and

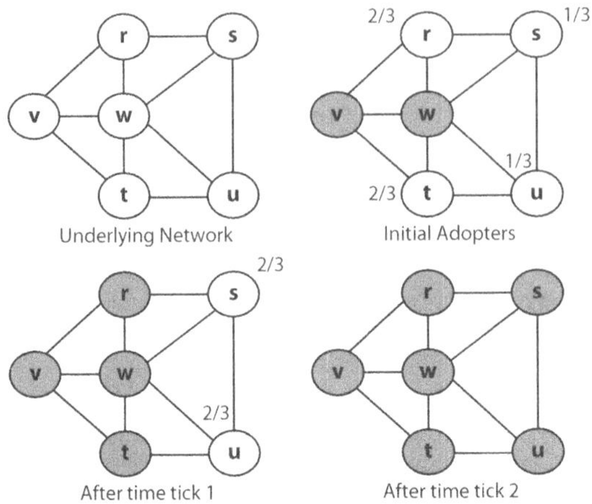

Figure 10.7 Cascading behavior.

B's coexistence. Assume that a = 3 and b = 2 in the coordination game; i.e., payoff to nodes that are interacting by using behavior A is 3/2 times that of nodes dealing with activity B. The threshold formula dictates that the nodes move from behavior B to behavior A when minimum $q = \frac{2}{3+2} = 2/5$ of their neighbors have opted A.

In Figure 10.7, assume that the nodes v and w are among the first to embrace behavior A, whereas everyone else opts for B. Beginning with the neighbors of the initial adopters, we see that this a sort of chain reaction where first r and t adopt the same behavior and then eventually s and u also adopt the same behavior. So the critical question that we have to answer is—What proportion of the neighbors should adopt a decision before a node starts adopting it? The reason why so much emphasis is given on the neighbors is because even in our everyday lives, we are influenced by the choices that the people around us make. For instance, if we have to adopt a new technology, we usually wait for people around us to start using that technology before we try our hands on them. Hence, the problem that we are trying to address here is that how many neighbors around us have to start using that technology (adopt a decision) for us to start using that as well [12].

Given some initial adopters, we will say a user will adopt the same if at least q fraction of his neighbor is adopting the same. These adoptions take place sequentially called as ticks or iterations as information about the option cascades through the network. The process is stopped when either all the users have adopted the same decision or if no user has adopted the decision in a particular iteration. What matters is to look into the largest value of this fraction, which would make it possible for the users in the network to adopt the same decision. This largest value q is called as the cascading capacity of a network.

Going back to our previous example where v and w are initial adopters and q has been assigned a value of 0.5. This means that each node will adopt the decision if at least half of its neighbors have adopted the same decision. Given these initial adopters, for each node we try to find the fraction of neighbors who have adopted the same decision. For node r, it has three neighbors, and two of its three neighbors have adopted the same decision, which is evident from the adoption value of r which is two thirds. For s, one third of its neighbors have adopted the same decision similarly for u, one third of its neighbors and for t, two thirds of its neighbors have adopted the same decision. Now, in every iteration, all such nodes will be considered, which have a value higher than 0.5. These nodes will turn green, that is, they will adopt the same decision. In iteration 1, we see that r and t have turned green, that is, they have adopted the same decision because they have a value of 2/3 which greater than 0.5. This changes the value of s and u to 2/3 because now two thirds of their neighbors have adopted the same decision. Thus, in iteration 2, s and u will adopt the same decision and turn green as now they have a value greater than 0.5. Now suppose if the value of q is 0.8, then in such a scenario none of the nodes will adopt the same decision, i.e., turn green because all the values are less than 0.8. This is because the cascading capacity of a network cannot be more than the largest or the maximum value of the nodes or the neighbors of the initial adopters.

So the largest value of q which facilitates the entire network to adopt the same decision is called the cascading capacity of the network. However, it is important to note that we cannot come up with the cascading capacity of a network without knowing the initial adopters, as well as the structure of the network, i.e., the distribution of the nodes and edges. Given the initial adopters v & w, cascading capacity of this network is 2/3.

To understand the cascading behavior using python code, let us consider a situation where every node has two options—either to choose behavior A or choose behavior B. There is an incentive for nodes v and w to have their behaviors match if they are connected by an edge. We depict that this with a game where the players are v and w. They have two possible strategies or decisions to choose from either strategy A or strategy B.

The following are the payoffs:

- Both the players will get a payoff a greater than zero if they adopt strategy A,

		w	
		A	B
v	A	a, a	0, 0
	B	0, 0	b, b

Figure 10.8 Payoff matrix in coordination game.

- Both the players will get a payoff b greater than zero if they adopt strategy B, and
- Both the players will get a payoff zero if they adopt the opposite strategies.

This can be written in the form of a payoff matrix, as depicted in Figure 10.8.

Figure 10.8 shows the situation on one edge of the network. However, every node v is mimicking the game with their neighbors. The payoff for v is the total of the payoff in games that are played on every edge. As a result, v's approach will be determined by the decisions made by every one of its neighbors collectively. Figure 10.9 shows the cascading payoff using python code. Codes for the key people or the active nodes are as shown in Figure 10.10. Impact of community on cascades and cascading on clusters are depicted in Figure 10.11 and 10.12 respectively.

10.11 Conclusion

Cascade networks are an example of networks that connect individuals on the basis of the direction in which the data or information flows between them. These networks have garnered the attention of various sociologists interested in the diffusion of innovation for many years; more recently, researchers in a variety of fields have looked into cascades for the purposes of identifying trendsetters for viral marketing, identifying inoculation targets in epidemiology, and also explaining trends in blog space. Research goals have shifted over time and across platforms, from simply seeing and counting cascades to tracking, anticipating flow of information, and modeling them. Despite extensive empirical work in sociology on data sets of modest size, data availability has reduced the scope of analysis. Understanding cascades is a crucial step in gaining a better understanding of how information spreads.

```
# cascade pay off
import networkx as nx
import matplotlib.pyplot as plt


def set_all_r(G):
  for i in G.nodes():
    G.nodes[i]['action'] = 'r'
  return G

def set_v(G, list1):
  for i in list1:
    G.nodes[i]['action'] = 'v'
  return G

def get_colors(G):
  color = []
  for i in G.nodes():
    if (G.nodes[i]['action'] == 'r'):
      color.append('red')
    else:
      color.append('blue')
  return color

def recalculate(G):
  dict1 = {}

  # payoff(V) = v = 1
  # payoff(R) = r = 5
  v = 15
  r = 5

  for i in G.nodes():
    neigh = G.neighbors(i)
```

Figure 10.9 Payoff in cascading behavior. *(Continued)*

```
    count_v = 0
    count_r = 0

    for j in neigh:
      if (G.nodes[j]['action'] == 'v'):
        count_v += 1
      else:
        count_r += 1
    payoff_v = v * count_v
    payoff_r = r * count_r

    if (payoff_v >= payoff_r):
      dict1[i] = 'v'
    else:
      dict1[i] = 'r'
  return dict1

def reset_node_attributes(G, action_dict):
  for i in action_dict:
    G.nodes[i]['action'] = action_dict[i]
  return G

def Calculate(G):
  terminate = True
  count = 0
  c = 0

  while (terminate and count < 10):
    count += 1

    # action_dict will hold a dictionary
    action_dict = recalculate(G)
    G = reset_node_attributes(G, action_dict)
    colors = get_colors(G)

    if (colors.count('red') == len(colors) or colors.count('green') == len(colors)):
```

Figure 10.9 (Continued) Payoff in cascading behavior. *(Continued)*

```
    terminate = False
    if (colors.count('green') == len(colors)):
      c = 1
  nx.draw(G, with_labels=1, node_color=colors, node_size=800)
  plt.show()
 if (c == 1):
  print('cascade complete')
 else:
  print('cascade incomplete')


G = nx.erdos_renyi_graph(10, 0.5)
nx.write_gml(G, "erdos_graph.gml")

G = nx.read_gml('erdos_graph.gml')
print(G.nodes())

G = set_all_r(G)

# initial adopters
list1 = ['2', '1', '3']
G = set_v(G, list1)
colors = get_colors(G)

nx.draw(G, with_labels=1, node_color=colors, node_size=800)
plt.show()

Calculate(G)
```

Figure 10.9 (Continued) Payoff in cascading behavior.

```
# cascade key people
import networkx as nx
import matplotlib.pyplot as plt


G = nx.erdos_renyi_graph(10, 0.5)
```

Figure 10.10 Keypeople in cascading behavior. *(Continued)*

```
nx.write_gml(G, "erdos_graph.gml")

def set_all_r(G):
  for i in G.nodes():
    G.nodes[i]['action'] = 'r'
  return G

def set_v(G, list1):
  for i in list1:
    G.nodes[i]['action'] = 'v'
  return G

def get_colors(G):
  color = []
  for i in G.nodes():
    if (G.nodes[i]['action'] == 'r'):
      color.append('red')
    else:
      color.append('green')
  return color


def recalculate(G):
  dict1 = {}

  # payoff(V)=v=4
  # payoff(R)=r=3
  v = 10
  r = 5
  for i in G.nodes():
    neigh = G.neighbors(i)
    count_v = 0
    count_r = 0

    for j in neigh:
      if (G.nodes[j]['action'] == 'v'):
        count_v += 1
```

Figure 10.10 (Continued) Keypeople in cascading behavior. *(Continued)*

```
      else:
        count_r += 1

    payoff_v = v * count_v
    payoff_r = r * count_r

    if (payoff_v >= payoff_r):
      dict1[i] = 'v'
    else:
      dict1[i] = 'r'

  return dict1


def reset_node_attributes(G, action_dict):

  for i in action_dict:
    G.nodes[i]['action'] = action_dict[i]
  return G


def Calculate(G):
  continuee = True
  count = 0
  c = 0

  while (continuee and count < 100):
    count += 1

    # action_dict will hold a dictionary
    action_dict = recalculate(G)
    G = reset_node_attributes(G, action_dict)
    colors = get_colors(G)

    if (colors.count('red') == len(colors) or colors.count('green') == len(colors)):
      continuee = False
      if (colors.count('green') == len(colors)):
```

Figure 10.10 (Continued) Keypeople in cascading behavior. *(Continued)*

```
      c = 1


  if (c == 1):
    print('cascade complete')
  else:
    print('cascade incomplete')


G = nx.read_gml('erdos_graph.gml')

for i in G.nodes():
  for j in G.nodes():
    if (i < j):
      list1 = []
      list1.append(i)
      list1.append(j)
      print(list1, ':', end="")

      G = set_all_r(G)
      G = set_v(G, list1)
      colors = get_colors(G)
      Calculate(G)
```

Figure 10.10 (Continued) Keypeople in cascading behavior.

```
# impact of communities on cascades
import networkx as nx
import random
import matplotlib.pyplot as plt


def first_community(G):
        for i in range(1, 11):
                G.add_node(i)
        for i in range(1, 11):
                for j in range(1, 11):
                        if (i < j):
```

Figure 10.11 Impact of communities on cascades. *(Continued)*

```
                r = random.random()
                if (r < 0.5):
                        G.add_edge(i, j)
    return G


def second_community(G):
    for i in range(11, 21):
        G.add_node(i)
    for i in range(11, 21):
        for j in range(11, 21):
            if (i < j):
                r = random.random()
                if (r < 0.5):
                        G.add_edge(i, j)
    return G



G = nx.Graph()
G = first_community(G)
G = second_community(G)
G.add_edge(5, 15)

nx.draw(G, with_labels=1)
plt.show()

nx.write_gml(G, "community.gml")
```

Figure 10.11 (Continued) Impact of communities on cascades.

```
# cascading on clusters
import networkx as nx
import matplotlib.pyplot as plt



def set_all_r(G):
  for i in G.nodes():
    G.nodes[i]['action'] = 'r'
```

Figure 10.12 Cascading on clusters. *(Continued)*

```
  return G


def set_v(G, list1):
  for i in list1:
    G.nodes[i]['action'] = 'v'
  return G


def get_colors(G):
  color = []
  for i in G.nodes():
    if (G.nodes[i]['action'] == 'r'):
      color.append('red')
    else:
      color.append('green')
  return color


def recalculate(G):
  dict1 = {}
  v = 3
  r = 2
  for i in G.nodes():
    neigh = G.neighbors(i)
    count_v = 0
    count_r = 0

    for j in neigh:
      if (G.nodes[j]['action'] == 'v'):
        count_v += 1
      else:
        count_r += 1
    payoff_v = v * count_v
    payoff_r = r * count_r

    if (payoff_v >= payoff_r):
      dict1[i] = 'v'
    else:
```

Figure 10.12 (Continued) Cascading on clusters. *(Continued)*

```
        dict1[i] = 'r'
    return dict1

def reset_node_attributes(G, action_dict):
    for i in action_dict:
        G.nodes[i]['action'] = action_dict[i]
    return G

def Calculate(G):
    terminate = True
    count = 0
    c = 0
    while (terminate and count < 100):
        count += 1

        # action_dict will hold a dictionary
        action_dict = recalculate(G)
        G = reset_node_attributes(G, action_dict)
        colors = get_colors(G)

        if (colors.count('red') == len(colors) or colors.count('green') == len(colors)):
            terminate = False
            if (colors.count('green') == len(colors)):
                c = 1

    if (c == 1):
        print('cascade complete')
    else:
        print('cascade incomplete')
    nx.draw(G, with_labels=1, node_color=colors, node_size=800)
    plt.show()


G = nx.Graph()
G.add_nodes_from(range(13))
G.add_edges_from(
    [(0, 1), (0, 6), (1, 2), (1, 8), (1, 12),
```

Figure 10.12 (Continued) Cascading on clusters. *(Continued)*

```
  (2, 9), (2, 12), (3, 4), (3, 9), (3, 12),
  (4, 5), (4, 12), (5, 6), (5, 10), (6, 8),
  (7, 8), (7, 9), (7, 10), (7, 11), (8, 9),
  (8, 10), (8, 11), (9, 10), (9, 11), (10, 11)])

list2 = [[0, 1, 2, 3], [0, 2, 3, 4], [1, 2, 3, 4],
   [2, 3, 4, 5], [3, 4, 5, 6], [4, 5, 6, 12],
   [2, 3, 4, 12], [0, 1, 2, 3, 4, 5],
   [0, 1, 2, 3, 4, 5, 6, 12]]

for list1 in list2:
  print(list1)
  G = set_all_r(G)

  G = set_v(G, list1)
  colors = get_colors(G)
  nx.draw(G, with_labels=1, node_color=colors, node_size=800)
  plt.show()

  Calculate(G)
```

Figure 10.12 (Continued) Cascading on clusters.

References

1. Freeman, L.C., Social network analysis: Definition and history, In *Encyclopedia of Psychology*, A. E. Kazdin (Ed.), American Psychological Association, Vol. 7, pp. 350–351, 2000.
2. Gross, J.L. and Yellen, J., *Handbook of graph theory*, CRC press, New York, 2003.
3. Ouyang, W., Wang, K., Zhu, X., Wang, X., Chained cascade network for object detection, in: *Proceedings of the IEEE International Conference on Computer Vision*, pp. 1938–1946, 2017.
4. Kermack, W.O. and McKendrick, A.G., Contributions to the mathematical theory of epidemics. II.—The problem of endemicity. *Proc. R. Soc. Lond. Ser. A*, containing papers Math. Phys. character, 138, 834, 55–83, 1932.
5. He, J., Zhang, S., Yang, M., Shan, Y., Huang, T., Bdcn: Bi-directional cascade network for perceptual edge detection. *IEEE Trans. Pattern Anal. Mach. Intell.*, Vol. 10, pp. 3828–3837, 2020.

6. Kramer, G., Yousefi, M.I., Kschischang, F.R., Upper bound on the capacity of a cascade of nonlinear and noisy channels, in: *2015 IEEE Information Theory Workshop (ITW)*, 2015, April, IEEE, pp. 1–4.
7. Shakarian, P., Bhatnagar, A., Aleali, A., Shaabani, E., Guo, R., The independent cascade and linear threshold models, in: *Diffusion in Social Networks*, pp. 35–48, Springer, Cham, 2015.
8. PM, K.R., Mohan, A., Srinivasa, K.G., *Practical Social Network Analysis with Python*, Springer International Publishing, New York, 2018.
9. Pathak, N., Banerjee, A., Srivastava, J., A generalized linear threshold model for multiple cascades, in: *2010 IEEE International Conference on Data Mining*, 2010, December, IEEE, pp. 965–970.
10. Kabir, K.A., Kuga, K., Tanimoto, J., Analysis of SIR epidemic model with information spreading of awareness. *Chaos, Solitons Fractals*, 119, 118–125, 2019.
11. Ghanbari, R., Jalili, M., Yu, X., Correlation of cascade failures and centrality measures in complex networks. *Future Gener. Comput. Syst.*, 83, 390–400, 2018.
12. Gouse, G.M., Haji, C.M., Saravanan, Improved reconfigurable based lightweight crypto algorithms for IoT based applications. *J. Adv. Res. Dyn. Control Syst.*, 10, 12, 186–193, 2018.

11

Exploring Social Networking Data Sets

Arulkumar N.[1]*, Joy Paulose[1], Mohammad Gouse Galety[2], Manimaran A.[3], S. Saravanan[4] and Saleem Raja A.[5]

[1]*Department of Computer Science, CHRIST (Deemed to be University), Bangalore, India*
[2]*Department of Information Technology, Catholic University, Erbil, Kurdistan Region, Iraq*
[3]*Department of Computer Applications, Madanapalle Institute of Technology and Science, Madanapalle, India*
[4]*Department of Electronics and Communication Engg., SRC, SASTRA University, Kumbakonam, India*
[5]*Department of IT, University of Technology and Applied Science-Shinas, Shinas, Sultanate of Oman*

Abstract
A network is a collection of objects/devices that are linked to one another through wired and wireless communication. Networks are everywhere, and they are formed to share resources among users. Nodes and edges form a network structure. Nodes describe the objects, whereas edges represent the connections between them. Network analysis is advantageous in a wide variety of live application activities. It enables us to get a thorough knowledge of the structure of a connection in social networks, the structure or process of development in environmental events, and even the study of organisms' biological systems. Additionally, network analysis allows the estimation of complex network patterns, and the network structure may be examined to disclose the network's basic features. If anyone examines a social connection among Facebook users, for example, the nodes indicate the target people and the edges reflect the connections between users, such as friendships or group memberships. The objective of this chapter is to describe and visualize social network analysis (SNA)

**Corresponding author*: itsprofarul@gmail.com

Mohammad Gouse Galety, Chiai Al Atroshi, Bunil Kumar Balabantaray and Sachi Nandan Mohanty (eds.) Social Network Analysis: Theory and Applications, (205–228) © 2022 Scrivener Publishing LLC

using Python and NetworkX, a Python framework for analyzing the structure, dynamics, and functions of complex networks.

***Keywords*:** Social network analysis (SNA), NetworkX, network patterns, network data sets

11.1 Introduction

Today, networks represent an important aspect of people's lives. Numerous real-world issues include relationships between data records. Networks have a significant impact on our daily lives, from providing valuable information to influencing elections. Networks allow network users to collaborate and share resources. It enables users to concentrate on the protection and processing of essential business information. This enables the network's different computers to get essential data from the central point. Different companies likely prefer to link their own computers, and so the need to organize computers into networks, so any computer on the network may interact with any other computer. Computer networking enables employees to cooperate more efficiently and easily share ideas. This increases their productivity and generates more money for the company. More importantly, computer networking improves how companies interact with the rest of the world. Logistics networks, the World Wide Web, the Internet, and social networks are all examples of network applications. Additionally, these types of graphs are often used in banking—for example, to illustrate financial transactions and the interconnectedness of central counterparties. Numerous features of graphs make them very useful for understanding the data they contain.

11.1.1 Network Theory

Network theory is the concept of networks as a description of either symmetric or asymmetric relationships among discrete objects. In computer science and network research, network theory is a branch of graph theory: a network may be represented as a graph having properties on its nodes and/or edges. It is important to realize that the aim of any kind of network analysis is to work with the complexity of the network to retrieve facts that would not be acquired by studying the functionalities separately. Here, the two basic components of the networks are nodes and edges [1].

A node is a location on a network that is linked to two or more other components. The nodes represent the things that will be examined, whereas the edges represent their relationships. Nodes may store both

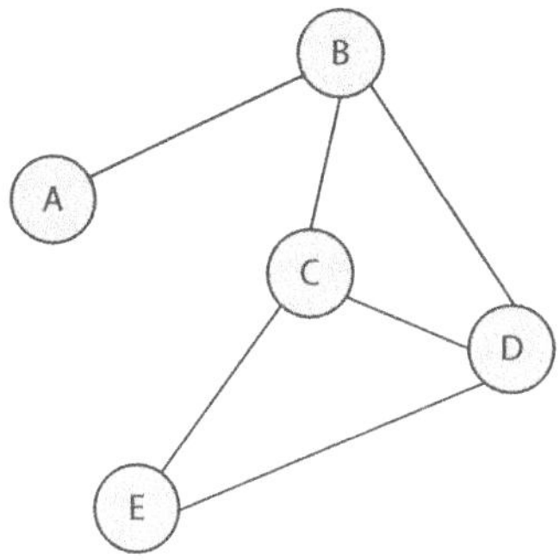

Figure 11.1 Nodes and edges in the graph.

self-referential characteristics (such as weight, size, location, and any other attribute) and network-referential information (such as degree-number, degree centrality, etc.). Edges represent the connections between nodes and may additionally contain characteristics, such as weight, which indicates the connection's strength, and direction. The nodes and edges in the graph are represented in Figure 11.1. Here, A, B, C, D, and E are the nodes, whereas the lines that connect these are referred to as edges.

11.1.2 Social Network Analysis

Network analysis is more like a study design that is well adapted for defining, researching, and analyzing different structural, as well as relational features. Network analysis is really a topological technique that emphasizes the patterns of actor-actor relationships. Social network analysis (SNA) is a method for studying social systems via the use of networks, as well as graph theory. The capacity to evaluate these networks and make excellent decisions is important for any data analyst. If analysts are analyzing a social link between Facebook users, for example, the nodes represent the target people and the edges indicate the connections between users, such as friendships or group memberships. It allows anybody to get a comprehensive understanding of the structure of a link in social networks, the structure or process of change in natural events, and even the biological systems of organisms [2]. Researchers are using quantitative network analysis in Python to browse between both the graph and the distinctive features of people and what ties them together [3].

In Figure 11.2, the circles represent the nodes and the lines connecting the circles are the edges. If this graph represented a social network, the circles would represent people, and an edge between two vertices could signify that those two individuals are friends.

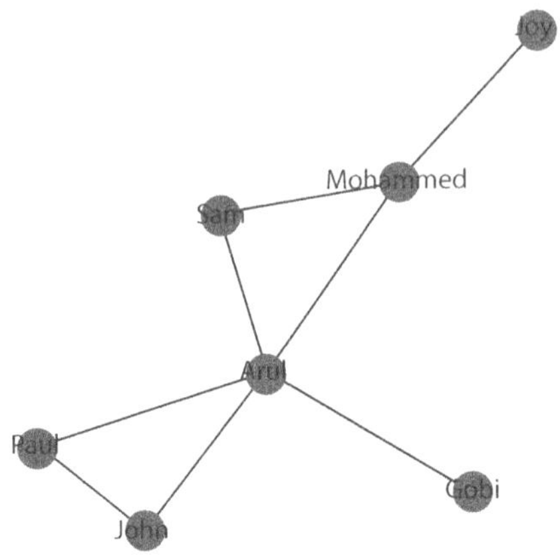

Figure 11.2 Nodes and edges in the social network.

- The term "node" refers to any kind of agent or element that we are attempting to communicate with. In this instance, it is individuals who are addressed. Here, Arul, Gobi, John, Joy, Mohammed, and Sam are the nodes.
- The edge is the link between two nodes. It is the physical interaction between individuals that is considered in the analysis. The link is the actual path that social network users are connected to. Here, the link which connects Arul and Sam is the edge of the network.

11.2 Establishing a Social Network

Social networks may be constructed from a number of data sets as long as the node-node connections can be specified. To convert the result to a tabular format, users can read the data from a data file (e.g., Excel) into a Pandas dataframe. Then, utilizing the edgelist in a pandas dataframe, developers can use NetworkX to build a directed graph. Finally, visualization techniques can be used to explore.

11.2.1 Designing the Symmetric Social Network

A symmetric social network is one in which all network traffic, including incoming and outgoing, is routed via a single channel. As a consequence, traffic flows into and out of the network along the same route.

A network is said to be symmetrical if its input impedance matches its output impedance. Physically symmetrical social networks are often, but not always, symmetrical networks. Occasionally, antimetrical networks are also important. These are networks with identical input and output impedances. In the symmetric social network, the connection of a symmetric network is simple; if user A is connected to user B, then user B is connected to user A [4].

In Figure 11.3, the Python program, which is used for creating the Symmetric network is presented. This symmetric network was created by using NetworkX in Python. In NetworkX, the Graph () method is utilized to create the network. An add edge is used to establish a connection between two nodes. In the given example, user A, user B, user C, user D, and user D are chosen as the nodes for creating the network. Here, if user A connected with user B, then user B connected with user A as well. In Figure 11.4, the output is received for the program given in Figure 11.3 and it is visualized in the form of a symmetric social network with five nodes.

```
import networkx as nx
G_symmetric = nx.Graph()
G_symmetric.add_edge('[illegible]','[illegible]')
G_symmetric.add_edge('[illegible]','[illegible]')
G_symmetric.add_edge('[illegible]','[illegible]')
G_symmetric.add_edge('[illegible]','[illegible]')
G_symmetric.add_edge('[illegible]','[illegible]')
G_symmetric.add_edge('[illegible]','[illegible]')
G_symmetric.add_edge('[illegible]','[illegible]')
G_symmetric.add_edge('[illegible]','[illegible]')
nx.draw_networkx(G_symmetric)
```

Figure 11.3 Python program for creating a symmetric social network using NetworkX.

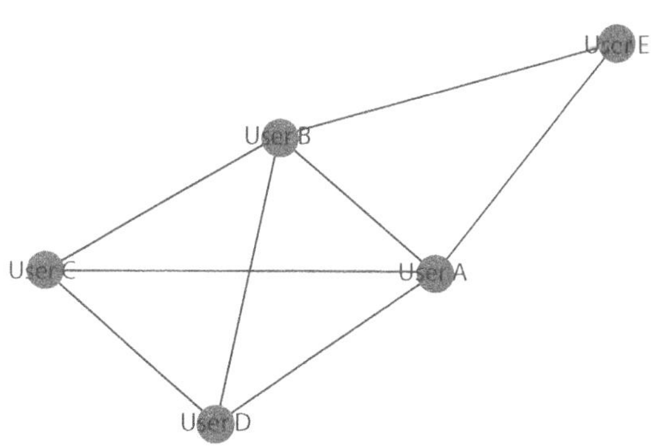

Figure 11.4 Visualizing the symmetric social network created by using NetworkX.

11.2.2 Creating an Asymmetric Social Network

An asymmetric social network has several routes for inbound and outbound network traffic. As a consequence, traffic enters and leaves the network along a distinct route. In computer networks, asymmetry refers to a substantial difference in the quantity of data or perhaps the speed of data flowing in one direction compared with the other during an average period. When symmetric access is utilized, the data transmission speed in both directions varies equally during an average period. The forward direction (from server to host) has a greater data transfer rate than the reverse direction (from host to server). The fundamental cause of other kinds of asymmetries in wireless networks, such as latency, is bandwidth asymmetry. When upstream packets take a different route than downstream packets, asymmetry in routing occurs. Wireless networks have higher packet error rates than wired networks, resulting in packet error rate asymmetry [5].

In SNA, asymmetric networks are ones in which nodes are not connected in such a manner that "if A is connected to B, B is connected to A." Consider the case of an "is the child of" connection. If A is not the child of B, then B is not the child of A. This network is asymmetrical. In NetworkX, an asymmetric network is created using the DiGraph or Directional Graph method.

In Figure 11.5, the given Python program is used for creating the Asymmetric network. This asymmetric social network was created by using NetworkX in Python. In NetworkX, the DiGraph () method is utilized to create the network. This DiGraph () method is an abbreviation for Directional Graph. Again, in the given example, user A, user B, user C, user D, and user D are chosen as the nodes for creating the asymmetric social network. In Figure 11.6, the graph is displayed as an asymmetric network for the program shown in Figure 11.5. As with the previous approach, the draw_networkx () function is used to visualize the network.

In Figure 11.7, the Python program creating the Asymmetric social network is included. Sometimes, nodes in the network depicted do not split

```
import networkx as nx
G_asymmetric = nx.DiGraph()
G_asymmetric.add_edge('User A','User B')
G_asymmetric.add_edge('User A','User D')
G_asymmetric.add_edge('User C','User A')
G_asymmetric.add_edge('User D','User E')
nx.draw_networkx(G_asymmetric)
```

Figure 11.5 Python program for creating the asymmetric social network using NetworkX.

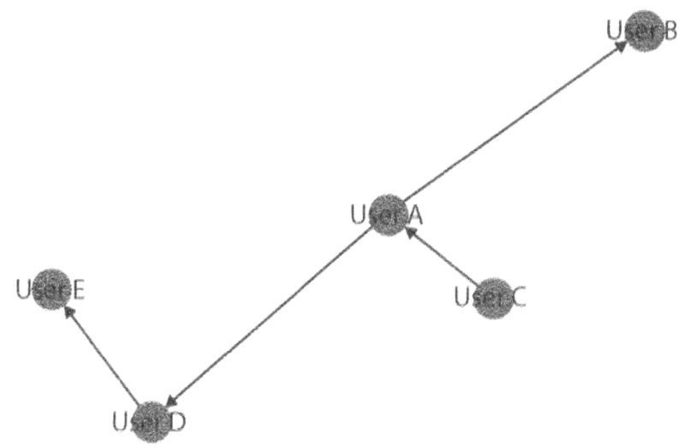

Figure 11.6 Visualizing the asymmetric social network created by using NetworkX.

```
import networkx as nx
G_asymmetric = nx.DiGraph()
G_asymmetric.add_edge([illegible],[illegible])
G_asymmetric.add_edge([illegible],[illegible])
G_asymmetric.add_edge([illegible],[illegible])
G_asymmetric.add_edge([illegible],[illegible])
nx.spring_layout(G_asymmetric)
nx.draw_networkx(G_asymmetric)
```

Figure 11.7 Python program for creating an asymmetric social network by applying the spring layout.

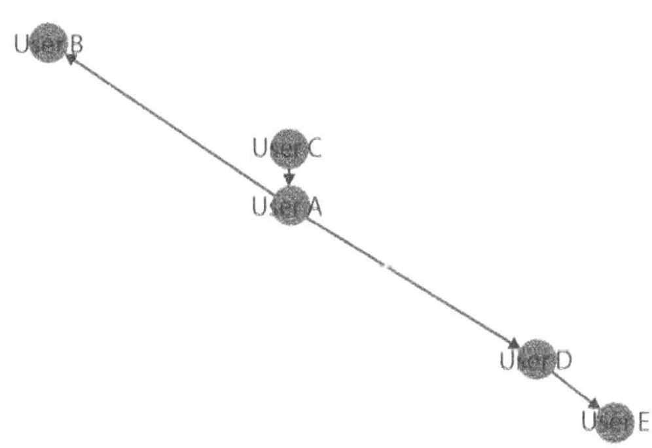

Figure 11.8 Visualizing the asymmetric social network created after applying the spring layout.

and are clearly visible. To address this, a layout function is imposed on node placement that enables us to view each node clearly. This visualization is done by utilizing spring_layout (). The improved social network is displayed as shown in Figure 11.8 by using the draw_networkx () method.

11.2.3 Implementing and Visualizing Weighted Social Networks

In a variety of real-world networks, not all connections in a network have the same capability. Weights are assigned to the connections between nodes in a weighted network. Weighted networks are also widely used in genomic and systems biologic applications. Varying the widths of the borders may also be used to indicate weights. Each edge of a weighted graph is assigned a numerical value (the weight). The value or weight of a weighted graph is determined by the sum of the edge weights throughout the cut.

In the previous examples, social networks are not assigned with any weights. In the case of the users, though, a weighted social network is created by giving the number of friends they have connected together with a weight. When a weighted social network is created, a weight should be provided to each edge, with the weight representing the number of projects or tasks which two users have connected to the network [6].

The example in Figure 11.9 shows the program to visualize the weighted social network created by applying the weights to the connections between

```
import networkx as nx
G_weighted = nx.Graph()
G_weighted.add_edge('User A','User B', weight=25)
G_weighted.add_edge('User A','User D', weight=8)
G_weighted.add_edge('User A','User E', weight=11)
G_weighted.add_edge('User A','User C', weight=1)
G_weighted.add_edge('User B','User D', weight=4)
G_weighted.add_edge('User B','User E',weight=7)
G_weighted.add_edge('User B','User C', weight=1)
G_weighted.add_edge('User C','User D',weight=1)
nx.circular_layout(G_weighted)
nx.draw_networkx(G_weighted)
```

Figure 11.9 Implementing the weighted social network using NetworkX.

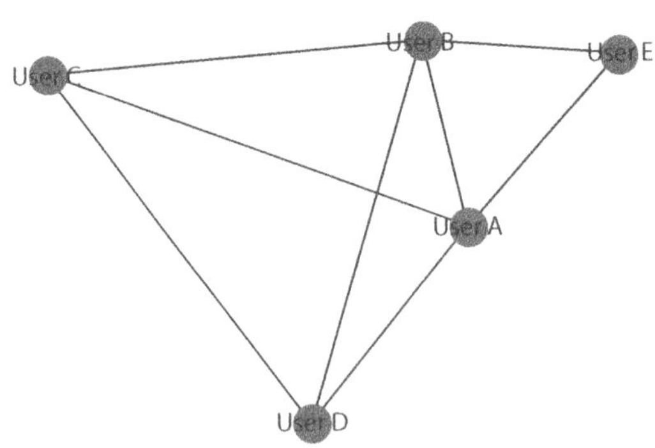

Figure 11.10 Displaying the weighted network in the form of circular architecture.

nodes in a weighted social network. The weight between two nodes in the network is specified by the edge width. Here, the weight = 25 is used for the connection between the node = user A and the node = user B. The weighted network of users in the form of a circular architecture is shown in Figure 11.10.

11.2.4 Developing the Multigraph for Social Networks

When many edges are allowed to connect any pair of vertices, the graph is known as a multigraph. In a multigraph, several edges may connect a pair of vertices. A multigraph is a graph with many properties on each edge. In this case, two nodes inside a social network may be connected through two different edges or connections. For instance, in addition to the existing connection (i.e., relation), the developer might establish a new one named "manager" between nodes A and B. The multigraph class for social networks has been used to construct a multigraph using NetworkX. A graph class that is undirected and may include a large number of edges. Multiedges is a term that refers to many edges between two nodes. Each edge may optionally hold data or attributes. A MultiGraph is a group of edges that are not directed [7].

This Python program in Figure 11.11 creates a graph for a social network, which is depicted in Figure 11.12 by implementing the Multigraph concept. Here, the relation keyword is used to connect users. Here, Manager, TeamLead, and Customer are the keywords chosen for assigning the relationships in the social network. Using the G_graph.edges () method, developers can verify the connectivity; the result would be as follows:

MultiEdgeDataView([('User A', 'User B', {'relation': 'Manager'}), ('User A', 'User B', {'relation': 'TeamLead'}), ('User B', 'User C', {'relation': 'Manager'}), ('User B', 'User E', {'relation': 'Customer'}), ('User C', 'User D', {'relation': 'TeamLead'})])

```
import networkx as nx
G_graph = nx.MultiGraph()
G_graph.add_edge([illegible],[illegible],relation =[illegible])
G_graph.add_edge([illegible],[illegible],relation=[illegible])
G_graph.add_edge([illegible],[illegible],relation=[illegible])
G_graph.add_edge([illegible],[illegible],relation=[illegible])
G_graph.add_edge([illegible],[illegible],relation=[illegible])
nx.draw_networkx(G_graph)
G_graph.edges(data=True)
```

Figure 11.11 Developing the multigraph for social networks using G_graph.edges ().

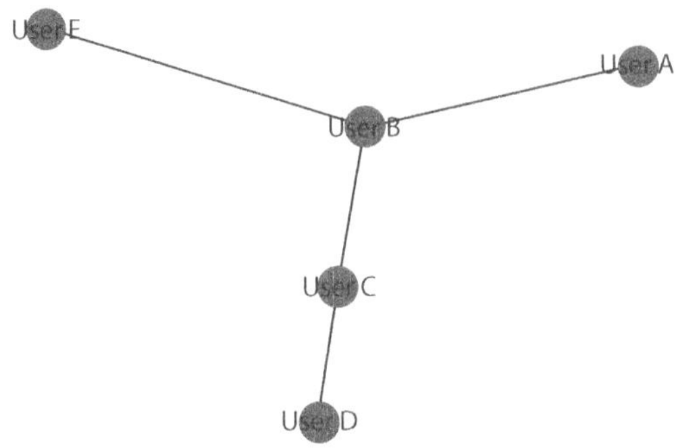

Figure 11.12 Results collected using the G_graph.edges () method.

11.3 Connectivity of Users in Social Networks

The topology of a network refers to the combination of its nodes and edges. Nodes are often known as vertices, whereas edges are the lines or arcs that link any two nodes in the network. The following are significant aspects that allow us to learn more about a certain node inside the network:

- The degree to which a network exists,
- Coefficient of clustering,
- The shortest route between two nodes,
- Eccentricity distribution of a node in a graph,
- Scale-independent networks,
- Transitivity.

11.3.1 The Degree to which a Network Exists

The degree of a node shows the number of connections it has. NetworkX provides a degree function that may be used to determine the degree of a node in a social network. A node's degree is defined as the number of edges that connect it. It is a fundamental parameter that has an impact on other properties, such as the centrality of a node. The distribution function of all nodes within the network helps determine if the networks are scale-free or not. In directed social networks, nodes have two degrees: an out-degree for edges that leave the node and an in-degree for those that enter [8, 9].

```
nx.degree(G_symmetric, 'User B')
```

```
nx.degree(G_graph, 'User B')

4
```

Figure 11.13 Degree of a node for symmetric social network.

The abovementioned given code will return a value of 4 because "user B" has only worked with four users on the network. The degree of a node is given in Figure 11.13.

11.3.2 Coefficient of Clustering

Individual users who share connections on a social network are seen to develop associations. In other words, a social network has a natural propensity to cluster. Developers can identify a node's clusters using the Local Clustering Coefficient, which is the percentage of pairs of a node's friends (that is, connections) that are linked in the network [10]. The nx.clustering (Graph_Type, Node_Name) function, which is shown in Figure 11.14, is utilized to calculate the local clustering coefficient.

According to the program and representation given in Figure 11.3 and Figure 11.4, user A has a local clustering coefficient of 0.6666666666666666. For the symmetric social network, the average clustering coefficient (total of all local clustering coefficients divided by the number of nodes) is 0.8666666666666666.

11.3.3 The Shortest Routes and Length Between Two Nodes

To represent the flow of information, the shortest paths, or the shortest distance in social between any two nodes, are utilized. The distance function

```
nx.average_clustering(G_symmetric)

0.8666666666666666

nx.clustering(G_symmetric, 'User A')

0.6666666666666666
```

Figure 11.14 Clustering and average clustering of a node for symmetric graph.

is used to calculate the gap between any nodes in a network. The shortest route is the one in which the fewest nodes are traversed. It finds the shortest cumulative impedance path between two nodes [11]. The route may connect just two points, the location, or it may include additional locations in a network.

By considering the program and representation given in Figure 11.3 and Figure 11.4, the shortest path is calculated. The nx.shortest_path(Graph_Type, 'Node1_Info', 'Node2_Info') function is considered to find the shortest path between any two nodes, and the nx.shortest_path_length (Graph, Node1, Node2) function is used to calculate the length of the path between any two nodes [12].

The output received in finding the shortest path between "user C" and "user E" in Figure 11.15 is ["user C," "user A," "user E"]. The result is received as 2 for calculating the shortest path length between the same nodes "user C" and "user E." In the same way, the value 2 is generated when computing the shortest path length between the nodes "user D" and "user E."

To calculate the distance between the node and every other node in the network, the breadth-first search technique is utilized by starting with one node. NetworkX provides the bfs_tree () function for this purpose. Thus, to try T = nx.bfs_tree (G_symmetric, "user C") and then show the tree, the graph given in Figure 11.16 describes how to reach further network nodes starting with user C. Furthermore, to execute T = nx.bfs_tree (G_symmetric, "user C") and then illustrate the graph, the diagram shown in Figure 11.16 illustrates how and when to reach further network nodes commencing with User C.

```
nx.shortest_path(G_symmetric, 'User C', 'User E')

['User C', 'User A', 'User E']

nx.shortest_path_length(G_symmetric, 'User C', 'User E')

2

nx.shortest_path(G_symmetric, 'User D', 'User E')

['User D', 'User A', 'User E']

nx.shortest_path_length(G_symmetric, 'User D', 'User E')

2
```

Figure 11.15 Calculating shortest path and shortest path length of a node.

11.3.4 Eccentricity Distribution of a Node in a Social Network

Eccentricity is determined by the length of the longest shortest path starting at a given node in a social network. It is also a term that refers to the greatest difference between one edge and all other edges in the network [13, 14]. This eccentricity distribution is represented by the letter e(V). In Figure 11.16, User C is chosen as the starting node, while in Figure 11.17, User A is chosen as the starting node for implementing the breadth-first

```
Tree_info = nx.bfs_tree(G_symmetric, 'User C')
nx.draw_networkx(Tree_Info)
```

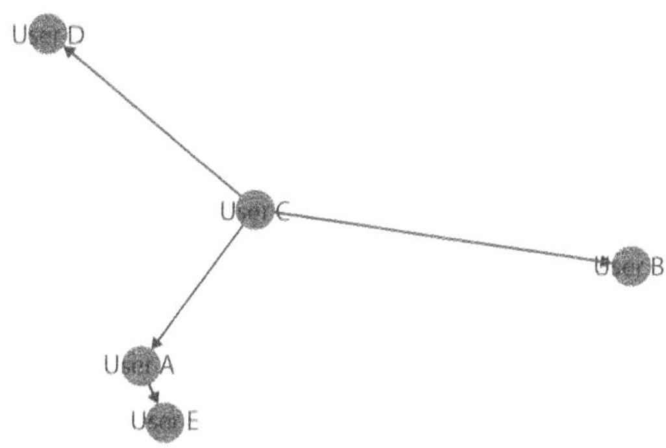

Figure 11.16 Implementing the breadth-first search algorithm for User C.

```
Tree_info2 = nx.bfs_tree(G_symmetric, 'User A')
nx.draw_networkx(Tree_info2)
```

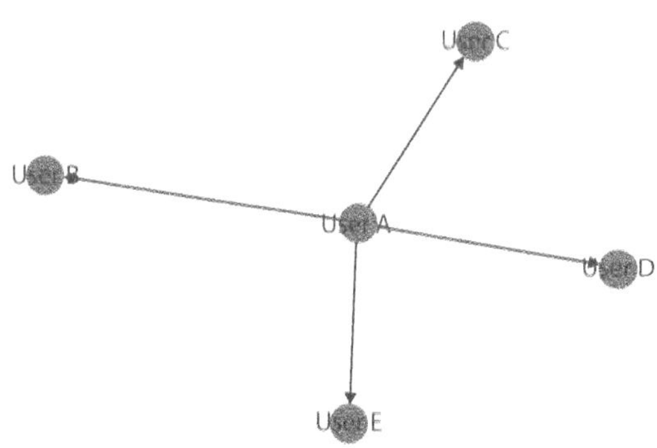

Figure 11.17 Implementing the breadth-first search algorithm for User A.

```
nx.eccentricity(G_symmetric,'User A')

1

nx.eccentricity(G_symmetric,'User C')

2
```

Figure 11.18 Eccentricity distribution of a node in a graph using the nx.eccentricity () function.

search. The nx.eccentricity (Graph_Type, 'Node_Info') function is considered to compute the value from the graph given in Figure 11.18. According to the network's eccentricity function, the node "user A" has an eccentricity of 1, whereas the node "user C" has an eccentricity of 2.

11.3.5 Scale-Independent Social Networks

Most nodes are connected to a small number of neighbours, but a few high-degree nodes provide the social network with a high degree of connectivity [15].

11.3.6 Transitivity

It refers to the presence of clusters or communities of nodes that are inextricably connected in a social network. These are node clusters that are more connected to one another than the rest of the network. Additionally, they are considered as topological clusters.

11.4 Centrality Measures in Social Networks

This is to ascertain the most important nodes in the social network. The term "centrality metrics" refers to these. Centrality measures can help us to find the most popular, most liked, and most influential nodes in the network. There are a few key concepts that help us gain more information about a particular node inside the social network [16, 17].

- Centrality by Degree
- Centrality of Eigenvectors
- The Centrality of Betweenness

The number of edges connecting a particular node is referred to as its degree of centrality. This might refer to an individual's friend count on a social media network. The centrality by degree is calculated using the degree () function.

According to graph given in Figure 11.17, the expression "nx.degree (G symmetric, "User A") is used to figure out the degree. The resulting output in this case is 4. User A is linked to the other nodes: User B, User C, User D, and User E.

11.4.1 Centrality by Degree

A node's degree indicates the number of connections it has. Individuals that are popular or liked often have the most friends on social networks. A node's degree of centrality in a network is a measure of its connectivity. It is based on the premise that important nodes have a high density of connections. NetworkX in Python provides a degree function that may be used to determine the degree of a node in a network [18]. Degree centrality () is a NetworkX function that returns the degree of centrality of each node in a network. According to the python program given in Figure 11.3, this results in a total of four, because user A has collaborated with just four of the network's users.

11.4.2 Centrality by Eigenvectors

Centrality by Eigenvectors in a social network is a centrality measure that assesses a user's centrality not only in terms of their connections, but also in terms of the centrality of their connections. As a result, eigenvector centrality may be important, and social networks and their study are gaining popularity at a rapid pace. Not the quantity of people with whom one is connected, but the kind of individuals with whom one is connected may be indicative of a node's importance [19].

It establishes the significance of a node based on its connections to other important nodes. The NetworkX function eigenvector_centrality () may be

```
nx.eigenvector_centrality(G_symmetric)

{'User A': 0.5100364187624349,
 'User B': 0.5100364187624349,
 'User C': 0.43904190094642953,
 'User D': 0.43904190094642953,
 'User E': 0.30693667343390046}
```

Figure 11.19 Centrality by Eigenvector using NetworkX () function.

used to calculate the eigenvector centrality of all nodes in a network and the result is given in Figure 11.19. Google's PageRank system is a variation of the Eigenvector centrality technique.

11.4.3 Centrality by Betweenness

Centrality by Betweenness or Betweenness centrality quantifies how often a node acts as a connector between two other nodes along the shortest path. It was created by Linton Freeman as a method to measure a human's impact on other people's communication inside a social network. It shows the frequency with which a spot occurs on the shortest path between two points. It quantifies how often a node occurs on the shortest path between two other nodes. Nodes with a high degree of betweenness centrality provide a substantial contribution to network communication/information flow. Nodes with a high degree of betweenness centrality have the potential to have strategic influence and control over others. An individual in such a prominent position may exert influence over the whole group by withholding or coloring facts during transmission [20, 21]. Betweenness centrality () is a NetworkX function that is used to determine the betweenness of a network and the result is given in Figure 11.20. It enables us to define whether or not to normalize betweenness data, whether or not to include endpoints in shortest route counts.

11.4.4 Closeness to All Other Nodes

The phrase "closeness to all other nodes or closeness centrality" refers to a node's capacity to transfer data across a network very efficiently. The closeness centrality of a node shows its average (inverse) distance to all other nodes. Nodes with a high proximity score have the shortest distances to all other nodes. Closeness-score-high nodes always have the shortest distance to all other nodes in the network. It is a helpful metric in social networks for

```
nx.betweenness_centrality(G_symmetric)

{'User A': 0.16666666666666666,
 'User B': 0.16666666666666666,
 'User C': 0.0,
 'User D': 0.0,
 'User E': 0.0}
```

Figure 11.20 Nodes with a high degree of betweenness centrality.

```
nx.closeness_centrality(G_symmetric)

{'User A': 1.0,
 'User B': 1.0,
 'User C': 0.8,
 'User D': 0.8,
 'User E': 0.6666666666666666}
```

Figure 11.21 Closeness to all other nodes is displayed for the G_symmetric graph.

estimating the speed with which information flows between two nodes [22, 23]. The result of the closeness_centrality () function is given in Figure 11.21.

11.5 Case Study of Facebook

To begin with the Facebook data; for this case study, the combined ego networks data set from Facebook was used, which has the aggregated network of ten people's Facebook friends list. The required combined.txt file for Facebook is obtained from the Stanford University website. For analyzing data, there are Facebook/Twitter APIs to retrieve your own Facebook/ Twitter data. This data set includes the "circles" (or "friends lists") on Facebook. This Facebook application was used to collect data from survey respondents. The data set contains node attributes (profiles), circles, and ego networks. By replacing a new value for every participant's internal Facebook id, the Facebook data have been anonymized. Furthermore, although feature vectors from this data set have already been made publicly accessible, their interpretation has now been concealed. Thus, using

```
import networkx as nx
import matplotlib.pyplot as plt
import pandas as pd
%matplotlib inline
import warnings; warnings.simplefilter('ignore')
```

```
df = pd.read_csv('/content/facebook_combined.txt')
```

Figure 11.22 Loading necessary packages and the dataset.

anonymized data, it is possible to determine if two people have the same affiliations, but not what those affiliations mean [24].

The complete case study analysis was done by using NetworkX in Python. The detailed analysis is as follows. In Figure 11.22, the necessary packages like NetworkX, matplotlib.pyplot and pandas are imported for the analysis. Pandas is a well-known Python-based data analysis toolbox that may be loaded using the "import pandas as pd" command [25, 26]. "read_csv" is a fundamental Pandas function for reading and manipulating text and csv files. This 'read_csv' function reads the facebook_combined.txt file.

The info () function in Python provides a concise summary of a data frame's contents. This method provides information on the index and column data types, non-null values, and memory usage of a DataFrame. This function df.info () displays the column, non-null, count and data type, as illustrated in Figure 11.23. According to the Python code given in Figure 11.24, the total number of nodes is 4039 and the number of edges in the data set is 88234.

Degree_centrality () function that returns the highest degree of centrality in the network. As per the Python program given in Figure 11.25, this degree of centrality results in a total of 107. Also, while using the nx.degree () function, it is clearly evident that "user 107" cooperated with just 1045 other network users.

```
df.info()

<class 'pandas.core.frame.DataFrame'>
RangeIndex: 88233 entries, 0 to 88232
Data columns (total 1 columns):
 #   Column  Non-Null Count  Dtype
---  ------  --------------  -----
 0   0 1     88233 non-null  object
dtypes: object(1)
memory usage: 689.4+ KB
```

Figure 11.23 Function info () to display the dataframe's contents.

```
G_fb = nx.read_edgelist("/content/facebook_combined.txt", create_using = nx.Graph(), nodetype=int)

print(nx.info(G_fb))

Graph with 4039 nodes and 88234 edges
```

Figure 11.24 Function info () to illustrate the nodes and edges in the data set.

```
dg_centrality = nx.degree_centrality(G_info)
sorted_dg_centrality = sorted(dg_centrality.items(), key=operator.itemgetter(1), reverse=True)
sorted_dg_centrality[:10]

[(107, 0.258791480931154),
 (1684, 0.1961367013372957),
 (1912, 0.18697374938088163),
 (3437, 0.13546310054482416),
 (0, 0.08593363051015354),
 (2543, 0.07280832095096582),
 (2347, 0.07206537890044576),
 (1888, 0.0629024269440317),
 (1800, 0.06067360079247152),
 (1663, 0.05819712729073784)]


nx.degree(G_info, [107])

DegreeView({107: 1045})
```

Figure 11.25 Degree_centrality () and nx.degree () functions.

```
print(nx.average_shortest_path_length(G_info))

3.6925068496963913
```

Figure 11.26 Average shortest path calculation between two networks.

The term 'average_shortest_path_length ()' refers to a notion in network topology that is defined as the average number of steps along the shortest route between any two network nodes. It is a metric that indicates the effectiveness of information or mass transmission on a network. The output of this metric is 3. 6925068496963913 and it is given in Figure 11.26.

The nx.draw_networkx () method is then applied (Figure 11.27) to display the Facebook data set as a graph (Figure 11.28).

To display the network in such a way that the color of the node changes with degree and the size of the nodes varies with betweenness centrality. The Python code for this betweenness_centrality () is described in Figure 11.29. The output of this code is presented as a graph in Figure 11.30 with different node colours.

The labels of nodes with the highest betweenness centrality are by using the sorted () formula. Besides, it displays the five node labels with their respective Betweenness Centrality values. Here, 107, 1684, 3437, 1912, and 1085 are the nodes with the highest betweenness centrality and they control the information flow in the network. It is normal for more connected nodes to be located on the shortest routes between other nodes. The node

```
plt.figure(figsize=(10,10))
nx.draw_networkx(G_info);
```

Figure 11.27 The draw_networkx () method to visualize the facebook data set.

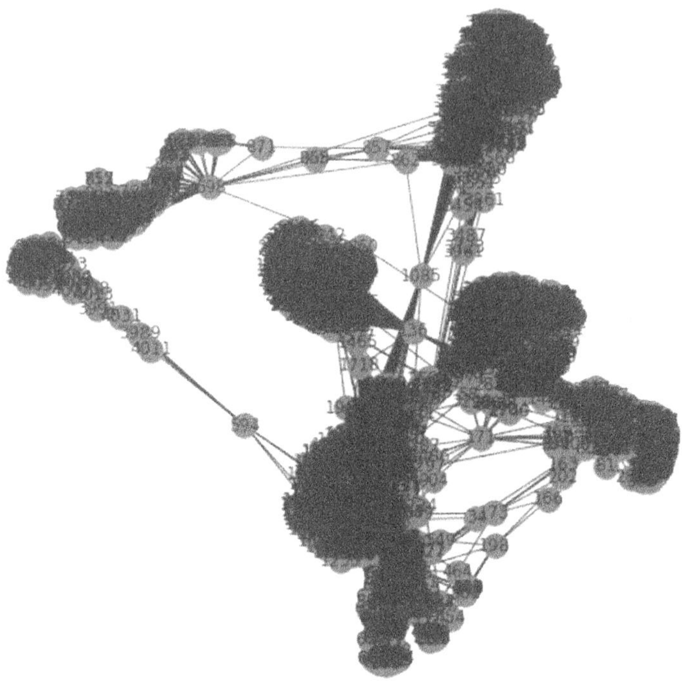

Figure 11.28 The visual representation of the facebook data set with draw_networkx ().

```
pos = nx.spring_layout(G_info)
betCent = nx.betweenness_centrality(G_info, normalized=True, endpoints=True)
node_color = [20000.0 * G_info.degree(v) for v in G_info]
node_size =  [v * 10000 for v in betCent.values()]
plt.figure(figsize=(10,10))
nx.draw_networkx(G_info, pos=pos, with_labels=False,
                 node_color=node_color,
                 node_size=node_size )
plt.axis('off');
```

Figure 11.29 Python code for betweenness_centrality ().

Figure 11.30 The visual representation of the data set with betweenness_centrality ().

```
sorted(betCent, key=betCent.get, reverse=True)[:5]

[107, 1684, 3437, 1912, 1085]
```

Figure 11.31 The sorted() method displays the nodes with the centrality.

```
g_fb_pr = nx.pagerank(G_info)
top = 10
max_pagerank = sorted(g_fb_pr.items(), key = lambda v: -v[1])[:top]
max_pagerank
```

Figure 11.32 PageRank() method to estimate popularity.

```
[(3437, 0.0076145868447496),
 (107, 0.006936420955866117),
 (1684, 0.006367162138306824),
 (0, 0.006289602618466542),
 (1912, 0.003876971600884498),
 (348, 0.002348096972780577),
 (686, 0.002219359259800019),
 (3980, 0.0021703235790099928),
 (414, 0.001800299047070226),
 (698, 0.0013171153138368812)]
```

Figure 11.33 Popularity nodes according to the page rank() method.

107 is significant since it is critical in the centrality metrics examined. This sorted () function in Python code is given in Figure 11.31.

The PageRank concept is used in Figure 11.32 to estimate the popularity of the incoming links in the network. The result of this PageRank () is given in Figure 11.33. It shows that the 3437 node is more famous than other nodes in the network.

11.6 Conclusion

This chapter discusses the significance of network analysis in a variety of areas and the NetworkX package's fundamentals. Additionally, a Facebook data set is explored that illustrate the use of network analysis. The purpose of SNA is discussed to get an overall understanding of an online community that maps the connections that link its members. It is identified that, using SNA, anyone can identify important people, groups within the network, and or associations between the groups. The key strategies in various social networking models, like Symmetric, Asymmetric, weighted, and multigraph, are designed using the NetworkX in Python program and the results are presented in graphical format. The topology of the networks was created to discover more about a particular user on a social network. The degree of a node indicates the number of users and their connections to other users in the network.

Local clustering and average clustering concepts are used to measure the degree to which nodes in a graph tend to cluster together. The shortest path was found between the users and it produces the distance between the users in the social network. The Eccentricity distribution of a node is defined in this context as the greatest distance between two adjacent edge pairs. The idea of centrality is used to identify the network's most important node, center node, and significant node. The idea of centrality has been used to identify the network's most important node, center node, and conspicuous node. This is achieved via the use of Degree Centrality, Eigenvector Centrality, and Betweenness Centrality. Finally, a data set from Facebook was analyzed to evaluate nodes with the greatest betweenness centrality and to assess their popularity using the Pagerank technique. The nodes 107, 1684, 3437, 1912, and 1085 have the greatest betweenness centrality, which regulates the network's data flow. The PageRank () function displays that the 3437 node is more popular than the rest of the network's nodes.

References

1. Neveu, A.R., A survey of network-based analysis and systemic risk measurement. *J. Econ. Interact. Coord.*, 13, 2, 241–281, 2018.
2. Dadpour, M., Shakeri, E., Nazari, A., Analysis of stakeholder concerns at different times of construction projects using Social network Analysis (SNA). *Int. J. Civ. Eng.*, 17, 11, 1715–1727, 2019.
3. Prabhakar, N. and Jani Anbarasi, L., Exploration of the global air transport network using social network analysis. *Soc. Netw. Anal. Min.*, 11, 1, 1–12, 2021.
4. Krishna Raj, P.M., Mohan, A., Srinivasa, K.G., *Practical Social Network Analysis with Python*, Springer International Publishing, Cham, 2018.
5. Goldenberg, D., Social Network Analysis: From Graph Theory to Applications with Python. In: *Proceedings of Israeli Python Conference 2019 (PyCon '19)*. New York, NY, USA, ACM, 2019.
6. Gotecha, M.R. and Patwardhan, M.S., Identification of key opinion leaders in healthcare domain using weighted Social Network Analysis, in: *2016 International Conference on Computing Communication Control and automation (ICCUBEA)*, IEEE, pp. 1–6, 2016.
7. Bhanodia, Kumar, P., Khamparia, A., Pandey, B., Prajapat, S., Online social network analysis, in: *Hidden Link Prediction in Stochastic Social Networks*, pp. 50–63, IGI Global, USA, 2019.
8. Ning, Z., Liu, Y., Zhang, J., Wang, X., Rising star forecasting based on social network analysis. *IEEE Access*, 5, 24229–24238, 2017.
9. Guzman, J.D., Deckro, R.F., Robbins, M.J., Morris, J.F., Ballester, N.A., An analytical comparison of social network measures. *IEEE Trans. Comput. Soc. Syst.*, 1, 1, 35–45, 2014.
10. Said, A., Abbasi, R.A., Maqbool, O., Daud, A., Aljohani, N.R., CC-GA: A clustering coefficient based genetic algorithm for detecting communities in social networks. *Appl. Soft Comput.*, 63, 59–70, 2018.
11. Swamynathan, G., Wilson, C., Boe, B., Almeroth, K., Zhao, B.Y., Do social networks improve e-commerce? A study on social marketplaces, in: *Proceedings of the first workshop on Online social networks*, pp. 1–6, 2008.
12. Pandia, M.K. and Bihari, A., Important author analysis in research professionals' relationship network based on social network analysis metrics, in: *Computational Intelligence in Data Mining*, vol. 3, pp. 185–194, Springer, New Delhi, 2015.
13. Akhtar, N., Javed, H., Sengar, G., Analysis of Facebook social network, in: *2013 5th International Conference and Computational Intelligence and Communication Networks*, IEEE, pp. 451–454, 2013.
14. Takes, F.W. and Kosters, W.A., Computing the eccentricity distribution of large graphs. *Algorithms*, 6, 1, 100–118, 2013.
15. Scarabaggio, P., Carli, R., Dotoli, M., A fast and effective algorithm for influence maximization in large-scale independent cascade networks, in: *2020 7th*

International Conference on Control, Decision and Information Technologies (CoDIT), Prague, Czech Republic, Vol. 1, pp. 639–644, IEEE, 2020.

16. Das, K., Samanta, S., Pal, M., Study on centrality measures in social networks: a survey. *Soc. Netw. Anal. Min.*, 8, 1, 1–11, 2018.
17. Landherr, A., Friedl, B., Heidemann, J., A critical review of centrality measures in social networks. *Bus. Inf. Syst. Eng.*, 2, 6, 371–385, 2010.
18. Zhang, J. and Luo, Y., Degree centrality, betweenness centrality, and closeness centrality in social network, in: *Proceedings of the 2017 2nd International Conference on Modelling, Simulation and Applied Mathematics (MSAM2017)*, vol. 132, pp. 300–303, 2017.
19. Maharani, W. and Gozali, A.A., Degree centrality and eigenvector centrality in twitter, in: *2014 8th international conference on telecommunication systems services and applications (TSSA)*, IEEE, pp. 1–5, 2014.
20. Kourtellis, N., Alahakoon, T., Simha, R., Iamnitchi, A., Tripathi, R., Identifying high betweenness centrality nodes in large social networks. *Soc. Netw. Anal. Min.*, 3, 4, 899–914, 2013.
21. Green, O., McColl, R., Bader, D.A., A fast algorithm for streaming betweenness centrality, in: *2012 International Conference on Privacy, Security, Risk and Trust and 2012 International Confernece on Social Computing*, Amsterdam, Netherlands, pp. 11–20, IEEE, 2012.
22. Yang, L., Qiao, Y., Liu, Z., Ma, J., Li, X., Identifying opinion leader nodes in online social networks with a new closeness evaluation algorithm. *Soft Comput.*, 22, 2, 453–464, 2018.
23. Khopkar, S.S., Nagi, R., Nikolaev, A.G., Bhembre, V., Efficient algorithms for incremental all pairs shortest paths, closeness and betweenness in social network analysis. *Soc. Netw. Anal. Min.*, 4, 1, 220, 2014.
24. Facebook dataset. (n.d.). SNAP: Network datasets: Social circles. Retrieved June 15, 2021, Website. https://snap.stanford.edu/data/egonets-Facebook.html
25. Software for complex networks. (2021, July 27). NetworkX: Network Analysis in Python. Retrieved November 29, 2021, Website. https://networkx.org/documentation/stable/index.html
26. Mohammed, G., Haji, C.M., Saravanan, Improved Reconfigurable based Lightweight Crypto Algorithms for IoT based Applications. *J. Adv. Res. Dyn. Control Syst.*, 10, 12, 186–193, 2018.

Index

Printed and bound by CPI Group (UK) Ltd, Croydon, CR0 4YY

17/09/2026

14964274-0002